# Painting in a World Transformed

William H. Libaw

# Painting in a World Transformed

*How Modern Art Reflects Our Conflicting Responses to Science and Change*

*by* William H. Libaw

Foreword by Michael Shermer

McFarland & Company, Inc., Publishers
*Jefferson, North Carolina, and London*

Library of Congress Cataloguing-in-Publication Data

Libaw, William H., 1923–
Painting in a world transformed : how modern art reflects our conflicting responses to science and change / by William H. Libaw ; foreword by Michael Shermer.
p. cm.
Includes bibliographical references and index.

**ISBN 0-7864-2211-4** (softcover : 50# alkaline paper) ∞

1. Painting, Modern—19th century. 2. Painting, Modern—20th century. 3. Art and science. I. Title.
ND1158.S36L53 2005
759.06—dc22 2005016990

British Library cataloguing data are available



On the cover: Gustava Caillebotte's *Rue de Paris, temps de pluie* (1877)
©2005 EyeWire, Inc.

Manufactured in the United States of America

*McFarland & Company, Inc., Publishers*
*Box 611, Jefferson, North Carolina 28640*
*www.mcfarlandpub.com*

# *Acknowledgments*

For looking at, commenting on, and otherwise helping with drafts of the manuscript or its parts, I am pleased to thank Cheryl Armon, Nina Barnett, Ellen Dinerman, Michael FitzGerald, Marsha Gaffery, Marsha Karioth, Jack Libaw, Oliver Libaw, Joseph Margolis, Nancy Mathews, Ellen Roberts, Michael Shermer, Jonathan Spaulding, Tim Spaulding, Carol Strickland, Bill Troffer, Linda Weintraub, Doug Wink, and Philip Yenawine. I am also indebted to many others, including the writers cited in the bibliography, for their views about art, artists, and other matters related to the cultures of both science and the humanities.

# Contents

# *List of Illustrations*

## *Foreword*

# *Art and Science in the Fullness of Life*

*by Michael Shermer**

In 1959 the physicist C.P. Snow famously divided the world of the mind into the "two cultures" of science and the humanities, with (as he saw it) an ever-widening gulf of communication between the two. Although Snow later confessed that there was nothing particularly original in his lecture (later expanded into a book)—intended primarily as a critique of the British public education system — it touched a cultural nerve that resonated for decades with (mostly) academics pontificating on how big the gap is, whether or not it is bridgeable, how it could be bridged, and what all is included in the catch-all category of "the humanities."

I must confess to being both fascinated by the power of Snow's model as well as frustrated by the murkiness of it. Obviously there is a difference between science and the humanities, and having a prominent scholar from one of the two camps clarify the issue helps put it on the cultural landscape as an issue worth addressing. But what precisely are the two cultures? Does "science and the humanities" include science and religion, science and history, science and literature, science and art? And why only two? Why is there not a "three-culture" problem of, say, science, religion, and art? Or science, history, and literature? Whatever component one plugs into the sec-

**Michael Shermer is the publisher of* Skeptic *magazine (www.skeptic.com), Executive Director of the Skeptics Society, a contributing editor and monthly columnist for* Scientific American, *and the author of* Why People Believe Weird Things, How We Believe, *and* The Science of Good and Evil.

ond half of the equation, the first is invariably science. There is a good reason for this that surely goes beyond the French structuralists' theory that the mind is wired to pigeonhole the world into binary categories (black and white, up and down, male and female, night and day). Science is not just different in some quantitative fashion from these other social and cultural traditions; it is qualitatively different. There is in science not just greater or fewer of elements found in these other ways of knowing; science has features nonexistent in any other knowledge tradition. Science is cumulative and progressive, with built-in self-correcting machinery to ensure that despite human foibles and fraud, the goal of approximating reality through data and theory has a reasonable chance of succeeding.

How science is defined, and what, precisely, should be included in a description of what science does, is a subject of considerable debate and disputation among philosophers and historians of science, with the current trend moving away from the extreme social-constructionist view of science proffered in the 1980s and 1990s, that argued science was, in effect, no different from art, literature, history, etc. This movement was an understandable backlash against the long-held positivist vision of science as an ever-improving approach to the Truth about Reality — an asymptotic curve of knowledge from the unknown to the known that, like all asymptotes, may never reach the upper ceiling of omniscience, but it nevertheless could come damn close. As a practicing scientist in the late 1970s (my field was experimental psychology — the name itself denotes a separation from those in the softer end of our field, clinical psychology) this was the view of science that I held. As I ground out countless hours in a behavioral lab studying the choice behaviors of rats, pigeons, and people (if you double the reward value for pressing the right bar instead of the left, will the number of bar presses on the right double those on the left?) I held fast to the conviction that as trivial as my research seemed in the larger scheme of things (I wasn't, after all, trying to explain how ordinarily good Germans became extraordinarily bad Nazis), I viewed each piece of data I collected as another brick in the edifice of science. There was Carl Sagan on PBS's *Cosmos* and Johnny Carson's *The Tonight Show*, pontificating on what it all means through the lens of science, and I was part of that tradition, however insignificant. It was a glorious canvas on which to paint!

A decade later I entered a doctoral program in the history of science, smack in the middle of that profession taking a "postmodern" turn toward reinterpreting science in a larger cultural context. Science is not an objective quest for the Truth about Reality, they averred; it is a subjective and socially constructed enterprise of knowledge building and power seeking by (mostly) white males seeking to justify the social status quo and cultural momentum of the Western tradition. For me it was, to use a more recent expression, a shock-and-awe experience. I was shocked to discover that, indeed, there is a very subjective side to science, that science is conducted by scientists loaded with the same cognitive biases and cultural prejudices as anyone else, and that whatever reality is, unraveling its secrets was not simply a matter of collecting more data because the facts never just speak for themselves. I was, however, awed by the observation that despite all these subjective components built into the fabric of my beloved profession, it still managed to construct so much objective knowledge; objective enough, in any case, to land spacecraft on distant planets, to decode the human genome, to unveil the workings of the brain, and to reveal the operations of an economy.

I admit to being a science junkie, while also confessing that my knowledge of the art world, like that of most scientists, is limited to flip-scans of coffee-table art books and whirlwind tours of the world's great art galleries and museums, audio headset firmly in hand. It's shameful, really, that too many of us scientists should be so insular in our perspective. Art is, after all, one of the first signs of our humanity, as defined by science itself! The magnificent cave paintings of our Paleolithic ancestors signify what paleo-anthropologists and archaeologists call "the great leap forward" in our evolution. Something big happened thirty-five to forty thousand years ago, and the artifact all scientists point to as proof of this leap is our artistic ability. Stone tools and cave paintings, science and art, objective and subjective, they are all inextricable.

This dualist theme of objective and subjective knowledge is brilliantly exposited by William Libaw in his explication in this book of one of the central themes of the two cultures war — science and art. Libaw describes another great leap forward, also represented in art, that began after science came of age. Thanks to science, the mirror is cracked and the visage of humanity, as reflected in our art, changed significantly when the great revolutions of science shattered our illusions of certainty and delusions of grandeur. Art, Libaw says, is subjective; it is "about life as it is felt, life as it is experienced by certain people in a specific time and particular place. Art may be about any of the countless inner activities which are felt in the living fabric of mental life." Science, by contrast, is objective; it "follows from reasoning and from information that is widely found useful and confirmed by others."

Yet subjective and objective cannot be unhinged. Who can look at a Hubble telescope photograph of a distant array of galaxies and not emote purely subjective feelings of transcendent wonder? Art, especially art in the age of science, reflects a very subjective reaction to very objective knowledge, particularly, says Libaw, "the effect of such knowledge and its consequences on our religious and spiritual understanding of our world and of ourselves." Libaw well captures this dualist balance between the objective and subjective, between the outer world and our inner experience of that world, whether that vision is expressed through the language of science or art. After reading this book you will never look the same way at a work of art from the past two centuries. I was amazed to see paintings of which I was familiar through a completely different light, and see in them what I had never noticed previously.

Consider the classic 1889 impressionist painting *The Starry Night* by Vincent van Gogh. It is a magnificent swirl of dark and light defined by stars, sky and land delineated by horizon, and the infinite vastness of space hovering over humanity's tiny abode (Figure 11-2). Van Gogh painted *The Starry Night* decades before Edwin Hubble expanded our universe by many orders of magnitude through his observations from the 100-inch Mt. Wilson telescope (where the universe extended from several thousand light years to many billions of light years across). But *The Starry Night* is the product of centuries of scientific discovery, coming after Copernicus displaced us from the center of the cosmos, after Kepler discovered the laws of planetary motion, after Galileo discovered the moons of Jupiter, mountains on the moon, and sunspots, after Newton united celestial and terrestrial physics, and after Darwin put us in our proper place in nature's complexity and diversity. No one, especially an emotionally volatile impressionist painter like van Gogh, could look up at the night sky and not be awed by the vastness of even the minuscule portion of the galaxy we can observe from earth.

Van Gogh painted the conflict between body and soul, between objective and subjective, and between outer and inner experiences. As he told his brother Theo: "I retain from nature a certain sequence and a certain correctness in placing my tones. I study nature so as not to do foolish things—however I don't mind so much whether my colour corresponds exactly, as long as it looks beautiful on the canvas." In Libaw's words, van Gogh "was one of the soldiers fighting to defend private experience and spiritual belief against the ever-expanding facts and information that were changing the world." Van Gogh himself, in fact, described *The Starry Night* "as an attempt to reach a religious viewpoint without God."

The same could be said of Carl Sagan's *Cosmos*. Science without art is dry and musty; art without science is hollow and superficial. We need both enterprises—the magisterium of science and the magisterium of art—not decoupled in some analytic formula, but united in the fullness of life.

# *Introduction*

Modern art can be about almost anything. The same is true of modern science. The two differ so greatly because, unlike what they can look at, the two peer from different places. Art looks at everything from the inside, the side of feelings, whereas science sees everything from the outside, the side of facts. As we will see, some modern art is about the outside world, which, to an ever-increasing extent, is full of the results of science and technology. Other kinds of modern art brush all that aside and focus on the inside world.

This book shows how modern art has reflected Western society's shifting feelings about the huge unending transformations in our world during the last two centuries, transformations that are primarily results of science and technology. However, the book is not a chronicle showing the development of modern art in itself, let alone a history of modern times. Neither, to be sure, is it about how artists have made use of science. Rather, it demonstrates how art has made images of our feelings and ideas about ever-changing science-based modern times as they have ranged between wonder and horror.

Art is shown to have reacted to the varying new reality with a widening separation between works where the artist turns more toward the outside to exhibit something about our shifting world, and creations that are increasingly infused with the artist's inside feelings. In so doing, modern art has also pictured the growth of Western society's fears that the material world might be overpowering the domain where the spirit dwells.

This new way of looking at recent art shows how major movements of the last two centuries are related to the greatest increase in knowledge in human history. The growth of science and technology provided material progress and improved health on a scale never seen before, yet science also gave us powerful new ways of endangering human life, and technology has been the source of endless artifacts that sometimes seem to overwhelm us with materialism. Painting has mirrored both sides of our reaction, reflecting our feelings that these changes were and are a blessing, and also a curse.

This theme is conveyed by the book's storytelling chapters, narratives illustrated with familiar works by well-known artists. Together, they show how we reacted to unceasing worldly change, either by being drawn more to the outside reality, or by turning inward to escape it. Each part is preceded by a brief chronology of significant developments in technology and science that were important to the era discussed in that part's chapters. Part I ("A Mirror with a Past") begins the book with brief chapters about the nature of our art, science, and knowledge. In Part II ("Scientism Begins: The First Half of the Nineteenth Century"), art is looked at in terms of Classical and Romantic painting. Pictures from Ingres and Delacroix are used to show the beginnings of what would become a division in art. In Part III ("The Changes Begin: The Second Half of the Nineteenth Century"), Impressionism is seen as an affirmative new way of looking at the rapidly changing world. This is shown with paintings from Manet, Monet and Renoir, and Morisot. A different reaction is found in works by Degas, van Gogh, and Gauguin, artists who created pictures that moved more toward the inner world of feeling, and Cèzanne, who created many works that depicted the indifferent outside world. In Part IV ("Science Repels and Attracts: The Early Twentieth Century"), the split widens with Matisse's decorative fascination with the inside world, and with Picasso's Cubism and Duchamp's common objects, both of which are oriented to the ever-novel outward side of life. Part V ("Looking Outside and Within: Early Mid-Century") shows two ways that painting responded to the blistering reality of the world in which we live. Hopper and Bacon made hard and harsh portraits of life in that world, while Abstract Expressionists Rothko, de Kooning, Krasner, and Pollock turned away and inward to picture subjective realities. Part VI ("Objects in the Mirror Are Seen as Shallow: Late Mid-Century") shows how Pop art and related styles practiced by Johns, Rauschenberg, Lichtenstein and Warhol turned back to the outside, but did so primarily to suggest a free-of-feeling world. Part VII ("Cracked but Not Broken: At the End of the Century") shows how, when mutually assured destruction of our civilized world no longer seemed likely, major in-between art returned with Richter, Kiefer, Fischl, Koons, and Salle as examples, although "extremist" art continued with Schnabel on the inside and Hirst on the outside. A concluding chapter sums up art's images of our concerns that the fruits of science and technology have been not only treasures, but also terrors, and even trash.

# PART I

# A Mirror with a Past

***Chronology of Technological and Scientific Developments***

5000 to 4000 B.C.E.: Egyptians utilize 360-day calendar
4000 to 3500 B.C.E.: Copper alloys used, gold and silver smelted
2500 to 2000 B.C.E.: Warfare aided by bow and arrow
800 to 700 B.C.E.: Spoked wheels and horseshoes employed
250 to 300 C.E.: Compass first employed
Around 1250: Recorded use of explosives
Around 1600: Telescope invented
Around 1620: Harvey discovers the circulation of blood
Around 1690: Newton's *Principia* uses mathematics to represent nature
Around 1740: Linnaeus classifies plants
Around 1770: Galvani discovers electrical nature of nervous impulse
Around 1775: Watt perfects the steam engine

## *Chapter 1*

# Soul or Machine?

To understand modern art, let's begin with two ideas.[1] The first is that art which we call modern is a true and vital expression of our own times. The second is that modern art is radically different from the art which preceded it.

Modern art is no joke, but some of it (such as the urinal named *Fountain* by Marcel Duchamp) is peculiar. Its mid-twentieth-century embodiments especially are nothing like the old kind of art. Some Abstract Expressionist art, such as Jackson Pollock's dripped sprinkles and splashes of paint, is less put together than a ghost. Some Pop art, such as Andy Warhol's multiple copies of soup cans, shows less spirit than does a zombie.

How did we get here, where art differs so much from what it once was? Art changed because the world changed, and we did too. Behind all the alterations in our world and ourselves was the huge continuing expansion of science. One of the results of science was that it filled the world with human-made objects. Also, science's study of humans hinted that we ourselves may be machine-like objects without souls or spirits within us.[2] It is all that change that shows up, indirectly but emphatically, in art.

Science began long ago; in fact, we can see something like science in what animals do to make their lives better. However, even as recently as the eighteenth century, science was still far from the top among our important ideas. What we call the Age of Science basically spanned the nineteenth century, loosely interpreted as running from the French Revolution to the First World War.[3]

Before the eighteenth and nineteenth centuries, religion not only dwarfed science (which was then regarded as the study of nature) but paid it little heed. Early in the seventeenth century, here's how a writer downplayed science: "We do not presume, by the contemplation of nature to attain to the mysteries of God.... Only let men beware ... that they do not unwisely mingle or confound these learnings together.... The more absurd and incredible any divine mystery is, the greater honor we do to God in believing it; and so much the more noble the victory of faith."[4]

For many who reacted to the eighteenth-century Enlightenment and to the nineteenth-century expansion of science, this mystification lost some of its appeal. At first, religion was not threatened by the new attitude that was developing. In fact, religion adopted science as an infant coming from itself: God's work was to be studied from two books, the good book of the Bible and the new book of nature.

> It became our duty as Christians to learn about nature, for natural religion was the cake on which revealed religion was the icing.... Newton was happy to see his discoveries turned to this use.... [However, eighteenth-century] high-churchmen ... were highly suspicious of the mechanical tenor of much science, and its tendency to make God a mere First Cause in what was called Deism.... [One new idea was the *Watch Argument*:] that if we should happen to come across a watch, and see how all the parts were made of suitable materials for the function they had to perform and how everything worked together, we should not take seriously the suggestion that its atoms happened by chance to have come together in this way.... [The argument] then seeks to demonstrate that the whole world is like a watch, and particularly that organisms are.[5]

Newton (1642–1727) was not only a genius, he was popularly perceived as such. Newton had shifted reason onto a new plane, and his rejection of dogma unsupported by fact became a model for an attack on the Church. In fact, it was Newton's spirit that fueled the extreme materialism expressed in a 1747 document where it is proposed that man himself is nothing but matter and motion.[6]

If we are merely matter in motion, that implies we are machines driven by indifferent cause-and-effect. Such ideas compete strongly with our widespread belief that it is our subjective feelings (that come from our spirits/souls/minds) that inspire us to make things happen. This conflict was a central source of stimulation to modern art — on the one hand, we began to worry about being mere machinery; on the other hand, we insisted that we really are embodied souls.

## *Chapter 2*

# What Is Art About?

Art is about life as it is felt, life as it is experienced by certain people in a specific time and particular place. What is expressed in art may be an intense emotion, an indefinite feeling of vitality or tranquility. Art may be about any of the countless inner activities which are felt in the living fabric of mental life.[1]

Art takes pictures of life as we conceive and experience our views of it. The prime function of art is "to make the felt tensions of life ... from the diffused [bodily stress] of vital sense to the highest intensities of mental and emotional experience ... stand still to be looked at."[2]

At this point, someone might well wonder: Doesn't that make *everything* the domain of art? The answer is a strong No! The domain of art is limited to the *subjective* side of life. That leaves out *objective* life and non-life, a large domain that is based on the organized facts that provide information about us and our world. As we will see, for some authorities it is objective life and non-life in the world that is everything. For them, subjective life is hardly worth bothering with; it may be no more than an idea that doesn't point to anything real.

Here is a related question and answer about art. What does the *artist* do? Whatever non-artistic interests (such as making a living or achieving fame) the artist may also have, the sole artistic intent is to embody an idea of some mode of feeling about something. That is what makes a person an artist. The product as such may or may not be good art, but it is art, whereas anything made with indifference to the feelings behind it is not art at all.[3]

What an artist does is *create* a work of art, not *construct* it, which is a dramatic way of saying the results must feel lifelike. For the artist, the simile "art must be like felt life" becomes the metaphor "art must be alive." This idea is portrayed literally in Jean-Léon Gérôme's painting *Pygmalion and Galatea*, 1881, which shows a sculptor working on a statue of an attractive young woman; it not only "comes to life," but rewards him with a kiss.

How we experience life varies from one person to another. It also varies from one epoch to another, which raises the question of the endurance and permanence of art. In *The Use and Abuse of Art*, Jacques Barzun pointedly addresses this issue by considering the grossly exaggerated remark, "'All else passes ... Art alone ... endures.' [Barzun then goes on to recall Ernest Hemingway's similar conceit.] 'A country finally erodes and the dust blows away, the people die and none of them were of any importance permanently, except those who practiced the arts.... A thousand years make economics silly and a work of art endures for ever.'" Barzun then punctures this overblown balloon: "Of course, the statement about lasting is false. *Some* art survives, partly by luck. And 'surviving' only means being valued and used by living people who are not artists."[4]

A bit of really old art now goes by the name *Polychrome Wall-painting of the Ice-Age in the Caves of Altamira, Northern Spain*, and was created around 15,000 B.C.E. Some artist or artists made images in those caves, bringing to life a pair of bison, one standing up and the other lying down and looking round. Of course, at the time they didn't perceive themselves to be artists, nor their creations to be art. It was "only in the middle of the eighteenth century in western Europe that concepts of art ... as we commonly employ them today began to develop."[5]

How could those ice-age people make art if they didn't know what it was? Clearly, the ability came into being long before the name. Likewise, chances are some people of that time were just as impressed as some of us are now with the occasional momentary brilliantly colored "pictures" in the sky that strongly convey to us the idea that the day is ending — we call them artistic sunsets (where the artist is Nature or God).

Barzun goes on to raise a controversial idea. He believes it is the *spirituality* of art that is its central characteristic. He writes, "But how can the spirituality of art be demonstrated? Quite simply by what we call its magic. Art proves the inadequacy of all materialistic explanations: in art the force and quality of the effect are out of all proportion to the cause.... If the devices of art were material only, they could be analyzed and reduced to formulas enabling any well-trained person to reproduce the magic at will...."[6]

In contrast with Barzun's idea, the view taken here is that it is the *subjectivity* of art's viewpoint, not its *spirituality* as such, that is its central characteristic. However, the whole area of human subjectivity and spirituality is a puzzle with little likelihood of wide agreement on how the two are related. Note that *both* spirituality and subjectivity are not generally regarded as *material*. Thus the realm of art is opposite to, or complementary to, the material terrain that is examined by science and other kinds of objective knowledge. The struggle between the subjective/spiritual realm and the objective/material domain is *itself* central to much of what artists of the last century or so have created for us to see.

## *Chapter 3*

# How Science Changed the Subject

The central idea of this book can be put like this: Art of the last century or so pictures our mixed reactions to (1) the vast expansion of science-related knowledge about things that range from the universe to our bodies, and (2) the effect of such knowledge and its consequences on our religious and spiritual understanding of our world and of ourselves.

Connected to this is the idea that we have two basic kinds of knowledge. "We may approach, enter into ... the thing known in order to attain subjective knowledge [for example, knowing where and how we hurt]. Or we may stand outside, observe ... the thing known in order to attain objective knowledge [such as what doctors use to help us]."[1]

Subjective knowledge is based on personal *experience*, on intuition, or on what we firmly believe with no clear evidence. Our knowledge of our own heartfelt religious beliefs (behind the behaviors, writings and institutions that have been constructed to represent them) is one kind of *subjective* knowledge. Objective knowledge follows from reasoning and from *information* that is widely found useful and confirmed by others. (Medicines to cure our maladies, books as such, and buildings to live in are some of the resulting products.)

Let's look first at objective knowledge. It includes not only science but also more down-to-earth things, like how to produce a pile of gadgets and then peddle them to a mass market. In a broad sense, objective knowledge — or, at least, objective information — was around long before we humans made our appearance. In fact, the "scientific method is not merely typical of all humans, but is also a key feature in the lives of most birds and mammals. Science [is] a highly formalized version of ... learning about regularities in the world."[2]

For an example of science without scientists, look at the use of medicinal plants by "Gelada baboons living in ... northern Ethiopia [who] face problems in the seasonal availability of their preferred foods (grass leaves). During the dry season, the grasses desiccate badly in the sun, and become very difficult to digest.... The animals become severely constipated. [However, near the dry season's end,] the wild roses come into fruit. The animals compete intensely for access to rosehips and consume them in large quantities.... [As a result, they] cease to suffer from constipation. [The] high vitamin C content [in rosehips makes them] an effective cure for constipation."[3]

Clearly, the baboons have useful objective information about the rosehips, where "objective" means something that is true for all of them. The word "information" is used here to suggest organized facts, whereas "knowledge" is an understanding of how the facts are interconnected. The baboons have some knowledge here, as they put these "medicines" in their mouths only when their guts play hardball with their rear ends.

Human beings (it hardly needs saying) have vastly more in the way of both objective information and knowledge than do baboons. The results from science have been behind most of the things we have made over the millennia. Anthropologist Ernest Gellner reminds us that you "cannot understand the human condition if you ignore or deny its total transformation by the success of the scientific revolution.... Our predicament is—to work out the social options of our affluent and disenchanted condition."[4]

Now let's switch over to look at subjective experience, belief, and knowledge. Most of it is certainly neither dramatic nor supernatural. Rather, it is the day-by-day experience of living with ourselves and with those we care about, and the beliefs and knowledge that we each have developed from that experience.

That said, let's shift to a pair of spectacular personal experiences, one hellish and the other heavenly. The first happened way back in 1844 to the father of William and Henry James, Henry James senior, when he "experienced something which would change [his and his family's] lives forever."

> He had eaten a good dinner. The children had been taken to bed. He was left alone, a rich, clever young man, staring into the embers of the grate. Quite suddenly and without warning James suffered a panic attack, and became aware that there was a presence in the room, "some damned shape squatting invisible to me within the precincts of the room and raying out from [its] fetid personality influences fatal to life." He wanted to run away from this

> terrifying demon or devil, but he could not. He remained rooted to the spot, suffering from "an evergrowing tempest of doubt, anxiety and despair, with absolutely no relief from any truth I had ever encountered save a most pale and distant glimmer of the divine existence...." This terrible experience was followed by two years of intermittent but profound depression.[5]

This ghastly experience may have happened only once to the elder James. However, it is somewhat like the events, called hypnagogic hallucinations, which occur at times to about ten or fifteen percent of people in the U.S.A. Such a person — who was in bed sleeping on his back — suddenly finds himself awake and paralyzed, and sees a terrifying creature approach him and then sit on his chest so that he can hardly breathe. These events are more subjective experiences than they are subjective knowledge. For some, they may become the basis of subjective belief that, say, Satan is messing around with their lives.

In contrast, here is a subjective experience, about as great as one can be, that improved the life of the man who had it. It happened to Richard Bucke, a doctor with a distinguished career. In 1872, at the age of 35, he and two friends had spent the evening reading. They parted at midnight, and he sat through a long drive in a hansom carriage. He tells us that his mind "was calm and peaceful.... All at once, without warning of any kind, [I found myself] wrapped around as it were by a flame-colored cloud. For an instant [I thought of] fire — some sudden conflagration in the great city; [the next instant, I knew] that the light was [within myself]. Directly afterwards came upon [me a sense] of exultation, of immense joyousness.... [Into my brain had] streamed one momentary lightning-flash [of splendor]."[6]

Bucke went on from this experience and developed a system of what here is called subjective belief. He might have called it knowledge, and he did call it Cosmic Consciousness. "[He believed this] consciousness shows the cosmos to consist not of dead matter governed by unconscious, rigid, and unintended law; it shows it on the contrary as entirely immaterial, entirely spiritual and entirely alive; it shows that death is an absurdity, that everyone and everything has eternal life; it shows that the universe is God and God is the universe, and that no evil ever did or ever will enter into it...."[7]

Such extraordinary experiences raise an issue about subjective knowledge that should be explicitly considered. That is: Is it true or is it false? The answer is both obvious and less than satisfying: *others* may not be able to gauge the validity of some of what *you personally* claim to know subjectively. You may know, for example, that your Uncle Bob is really nothing but a big selfish butt-head. You may know that for the rest of your life, or until—to your utter astonishment—you have experience of him that suggests that you have been just plain wrong all along.

That brings up a related question, one that also has an answer that is as plain as it is distasteful. Can we rely on *objective* knowledge to be rigorously valid? The main reasons we regard such knowledge as reliable and valid are: (1) It is agreed upon by many people, and (2) it works for them—they find it useful. Look then, for instance, at African healing practices, which rely on interpretations of visions, the summoning of ancestral spirits, and the consumption of herbal concoctions. There are hundreds of thousands of such traditional healers, and even more of their patients, who believe that such practices do indeed heal. (Never mind that there are also lots of healers—practitioners of modern medicine—who don't believe that.) If such traditional knowledge and practice were indeed totally valid, surely it would have spread out from Africa to the rest of the world. But it hasn't done that. Also, we in the Western world now know that both a doctor's attention and his medication (even if it is only empty-of-medicine placebos) often work, in the sense that the patient feels better.

*Total* rigor about the truth of knowledge is just not something that is available to human beings. However, even if some of it proves short of certain truth, the ideas of *subjective knowledge* and *objective knowledge* will help in making sense of modern art.

Although they have been discussed separately, subjective experience and objective information are mostly interwoven and entwined in the course of our normal activities. Indeed, the two are often the basis for internal discord. Consider the conflict implicit in the remark, "Old age is not for sissies." Those words come from movie star Bette Davis, who spoke them near the end of a long life. The comment surely suggests her knowledge that thinking about her own declining abilities and approaching death was frightening (this was her subjective knowledge). The assertion also shows her intent to get on with life despite dwindling competence and inescapable mortality (such objective knowledge—that death is inevitable—is possessed by all humans save the very young).

René Descartes' famous statement, "I think, therefore I am," also illustrates subjective knowledge and objective knowledge. His *thinking* is something of which he alone had (subjective) experience and knowledge. He used that knowledge to assert what everyone around him knew objectively: the man *existed*. Before he issued that statement, Descartes took pains to try to free his mind from false opinions about human nature. Then he followed it with a couple of assertions: "[My] whole nature or essence [is] only to think." "To exist, [that essence] has no need of space nor of any material thing.... Even if the body were not [to exist], [that essence] would not cease to be all that it is now."

There is something worth noticing about these last assertions of Descartes. Clearly, he places a far greater value on subjective mind than on objective body. Much modern religion does something similar — it emphasizes that it is the (subjective) soul that is to be saved, and provides only marginal hope that the body might be not saved but reconstituted or replaced. Ancient religions called Gnosticism and Manichaeism were likewise not only dualistic but both had a similar preference. For them, it was the *spiritual* kingdom of Light that was good, and the *material* kingdom of Darkness that was bad. A similar preference was present in ancient Greek philosophy. Plato suggested "the existence of two universal principles, Idea and Matter.... [Immaterial ideas] were eternal and immutable ... with perfect existence in every respect ... [whereas the] material world of sensible objects ... was always an imperfect imitation [of the world of Idea.]"[8]

Such a widespread and ancient view suggests two things. First, all of us — of whatever stripe and conviction — have *separate*, and sometimes competing, compartments in our heads for the two kinds of knowledge. Second and more important here, most of us *favor the subjective side* of that mental street. Experience (based on mind/soul/spirit) is usually more meaningful to us, more highly esteemed, and more likely to move us emotionally than is information (which is based on material/body/brain). For a petty example, look at how TV commercials supply no more than the bare minimum of necessary information. That allows almost all of their time to be filled with performers enacting experience that will get the viewer engaged with the product.

Although objective information and knowledge have been growing since our species came to life, the rate of such growth began its accelerated expansion only a few centuries ago. Now, we and our libraries struggle to contain the results. Sociobiologist Edward O. Wilson says, "no intellectual vision is more important and daunting than that of objective truth based on scientific understanding.... It took modern form in the eighteenth century Enlightenment hope that science would find the laws governing all physical existence.... The Enlightenment dream faded before the allure of Romanticism; ... more important, science could not deliver in the domain most crucial ... the physical basis of mind."[9] That some scientists are still trying to move our minds across the street is evidenced by the very title of a 2000 book by Edelman and Tononi: *A Universe of Consciousness: How Matter Becomes Imagination.*[10]

The world-changing expansion of objective science-like knowledge is behind much of our anxiety about our subjective (often religious) beliefs and knowledge. That anxiety and perceived threat is a large part of what modern art deals with. However, this book is not a history of that concern. It is more like the frames of a silent movie — a set of portraits of us taken by art over time — with some voiceover words.

# PART II

# Scientism Begins: The First Half of the Nineteenth Century

### *Chronology of Technological and Scientific Developments*

1800: Volta produces first battery

1810: Technique developed for canning foods

1827: Photographs produced on a metal plate

1829: Early version of the electromagnetic motor constructed

1831: Chloroform invented

1831: Maxwell suggests light and electromagnetism have identical source

1833: Gauss and Weber devise the electromagnetic telegraph

1834: Babbage invents the principle of the modern computer

1839: Goodyear formulates vulcanized rubber

1845: First submarine cable laid across English Channel

1846: Sewing machine patented

1849: Speed of light first measured

## *Chapter 4*

# The Handwriting on the Gallery Wall: Classical vs. Romantic

The nineteenth century was the start of the rapid growth of science, which was then thought of as the study of nature. Optimism that came with the many real results of science also contributed to scientism, which is the notion that science not only is the road to all useful reasoning but that it will develop answers to all reasonable questions.[1]

We now think of science as the second of what are called our "two cultures," where the older first culture consists of literature, languages, arts, philosophy, and history. Back then, the two cultures were still together. That was so in good part because the establishment's underrating of science made it still confident that the study of that "book of nature" was similar to studying the Bible. After all, both were God's books. However, even if there were no early-nineteenth-century breaks in the walls of the temple of unified belief and knowledge, there was something new showing on those walls, something that could be seen as handwriting, or as cracks induced by earthshaking science.

In *The Children of Frankenstein* we read that "The triumph of Newton led to an oddly contrary extreme: physics became the dominant science.... Men were dazzled by his vision of a clockwork universe, God as the Clockmaker who had wound it up for all time.... [But] why should men be pleased to live in a world conceived as a machine? It would seem to be a poor habitation for the human spirit, no less if this was called an immortal soul, for how this spirit could be free in a machine order was not at all clear."[2]

Those who reduced God to no more than a clockmaker were trying to paper over some of those suspicious marks on the walls in the temple. But that covering also contributed to the cracks underneath. Not only did a machine-universe make a poor habitat for soulful spiritual humans, it raised the painful possibility that God Himself no longer mattered much, whereas it began to seem that machines mattered too much. For example, in the 1818 book *Frankenstein*, the monster that was brought to life was a machine-like artifact that overran the power of its human creator to control it.

The science-religion struggle was related to a similar conflict between the Enlightenment and the Romantic movement that followed it. "The Romantic movement [developed] as a reaction to the rationalism that had come to dominate academic and public thinking during the Enlightenment of the preceding century. Its driving force lay in an attempt to reinstate the emotions and the immediacy of sensory experience in the forefront of human existence. Its main thrust was self-consciously anti-science, emphasizing the poetical and the spiritual over the empirical and the rational.... [Many in the Romantic movement sought] a purpose for life in a universe that contemporary astronomy and biology were inexorably demonstrating to be both accidental in origin and purposeless in existence."[3]

What is represented in art has two poles, one pointing inward and the other outward. Romanticism and religion are centered on the subjective pole of the pair that was behind art's cracking apart. The other pole, the objective one, is central to the Enlightenment confidence in reason, and to science's power to change the world.

In the art world of the early nineteenth century, Jean-Dominique Ingres was regarded as a Classical painter. "Classical" not only refers to ancient Greece and Rome; it also emphasizes rationality, which puts Ingres on the *objective* side of what would become a split in art. In that art world, Eugène Delacroix was not only a relative newcomer, but was regarded as a Romantic painter. So he is on the *subjective* side of that division. Looking backward, however, the beginnings of a divided art gallery are barely visible. It shows up mainly as more emotion in Delacroix's paintings than in those of Ingres, as we'll see in the following chapter.

## *Chapter 5*

# The Line Is Drawn: Ingres vs. Delacroix

Look at the way we were with art before it became modern. A painting created by Jean-Dominique Ingres called *Agamemnon sends his messengers to Achilles to persuade him to return to the fight*, 1801 (Figure 5-1) launched the young Ingres on his career as a history painter, where *history* painting is often *story* painting.

In this case, Ingres' story comes from the ancient *Iliad*. Remember that Agamemnon (the leader of the Greek invaders of Troy) had to give up his war-booty woman, Chryseis. In turn, Agamemnon used his power to grab Achilles' new girlfriend, Briseis. Achilles reacted by sulking in his tent. Politician Agamemnon then sent wily Odysseus and others to talk Achilles out of his snit. In Ingres' picture, we see Achilles seated on the left (next to his standing friend Patroclus), as cloaked Odysseus coaxes him: "Hey, let's make a deal. I'll remind Aggy that you're like his main man. But you gotta lighten up and come out of your tepee. Shake?"

Our two-hundred-years-later attention, however, focuses on what now seems odd. Most of the figures are wearing no clothing. That *dates* the picture for us, removing some of the value it was meant to have. Unclothed males were a *convention* in history pictures two centuries ago. They were also conventional in the time of the ancient Greeks, at least when they were engaged in athletic contests.

Many of our ways of doing things start out with a reason. Then, as people lose conviction in that basis and fail to keep track of it, those ways become merely conventional. In pages to come, we'll see many new ways of representing things in painting, ways which lost their impact and became

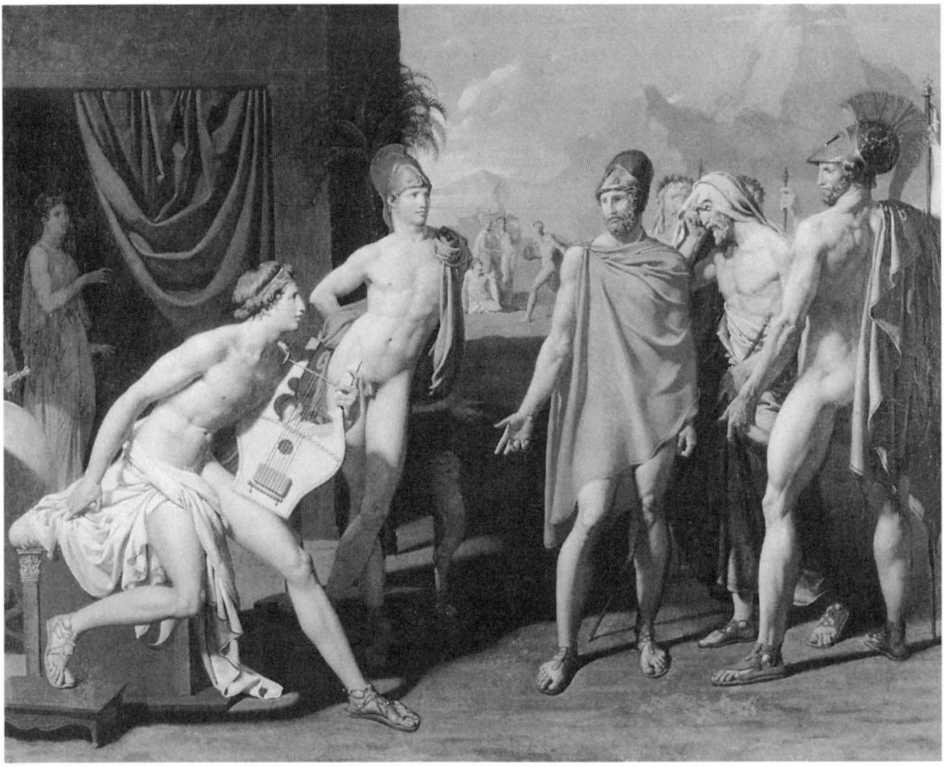

FIGURE 5.1. **Jean Auguste Dominique Ingres (1780–1867).** ***Agamemnon sends his messengers to Achilles to persuade him to return to the fight*****, 1801. Ecole des Beaux Arts, Paris, France.** PHOTOGRAPH COURTESY OF ERICH LESSING/ART RESOURCE, NY.

conventional after a while. Apart from nudity itself, there is another convention visible, one about masculinity, in Ingres' picture.

The author of *Male Trouble: A Crisis in Representation* tells us about it: "One reason the picture was so well received ... was its assured demonstration of total mastery over the recognized codes of masculine beauty. At least three distinct body types are represented ... the more muscular, mature and articulated bodies of Ajax and Odysseus ... the brooding Phoenix ... his body recognizably older than the others' ... [and] Achilles and Patroclus, [who are] smoother, lissome, younger, graceful, less muscular and conspicuously whiter-skinned.... But it is Patroclus's strikingly exaggerated outflung hip ... that signals one of the forms by which erotic differentiation could be incarnated in male bodies.... [It suggests] androgynous eroticism...."[1]

Be that as it may, there's no reason to project into the two-centuries-

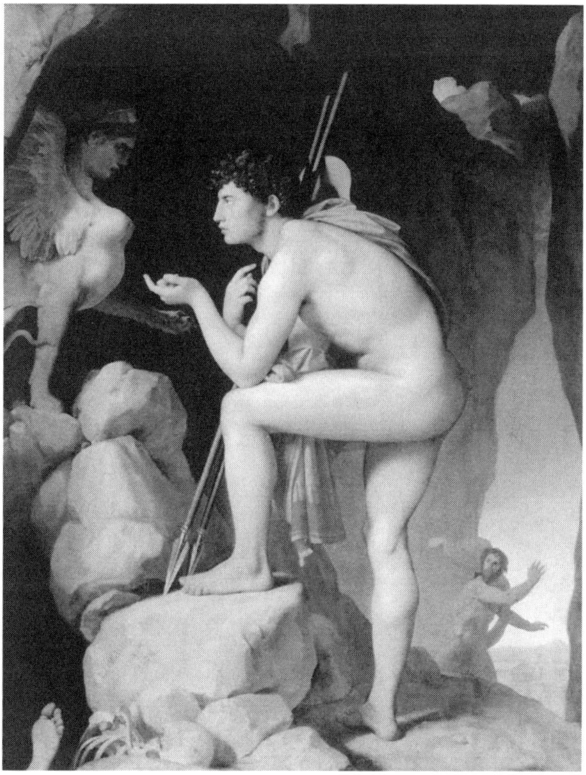

FIGURE 5.2. **Jean Auguste Dominique Ingres (1780–1867). *Oedipus explains the riddle of the Sphinx*, 1808. Oil on canvas. Louvre, Paris, France. PHOTOGRAPH BY R.G. OJEDA. PHOTOGRAPH COURTESY OF REUNION DES MUSÉES NATIONAUX/ART RESOURCE, NY.**

old picture one of the notions of our own time, namely, that bisexual men are really *homosexual* men. Patroclus and Achilles may have gone both ways, but they were not (remember, all this fuss was about Achilles' girlfriend, Briseis) what we sometimes call queer. For that matter, although an "exaggerated outflung hip" may suggest male effeminacy or homosexuality to us, it may not have implied much at all to the ancient people of Achilles' time. Their conceptualization of sex roles was not like ours. They (or at least the Greek ruling class) saw one partner in a sexual pair as powerful, male, and active, with the other partner as a female or a less important male and relatively passive. In *The History of Sexuality*, Foucault tells us they viewed sexual behavior as part of "the relationship between a superior and a subordinate, an individual who dominates and one who is dominated...."[2]

A few years after painting the *Iliad* scene, Ingres painted another fun story. This one is more mythical than historical: *Oedipus explains the riddle of the Sphinx*, 1808 (Figure 5-2). Nonetheless, every detail of bare-butted him and double-breasted her is painted as realistically as a perfectly focused photograph. (In fact, it wasn't until the last couple of decades of the nineteenth century that photography started to be widely used. By the time that occurred, painters like van Gogh and Gauguin no longer sought realism with precise representation in their paintings.)

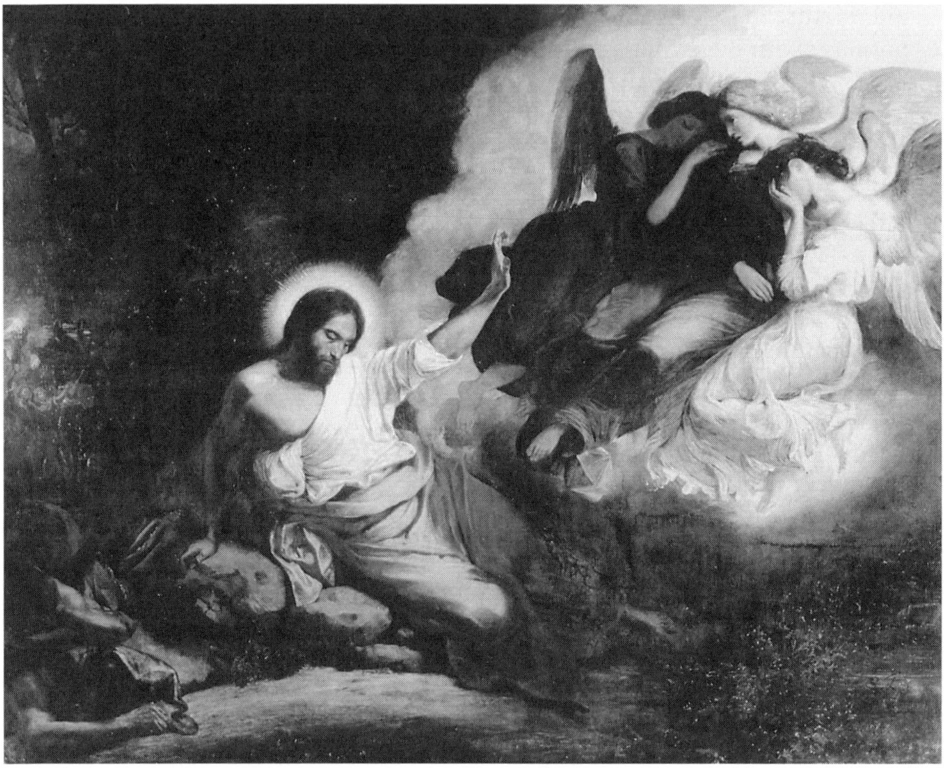

FIGURE 5.3. **Eugène Delacroix (1798–1863). *Christ in the Garden of Olives*, 1827. St. Paul– St. Louis, Paris, France.** PHOTOGRAPH COURTESY OF BRIDGEMAN-GIRAUDON/ART RESOURCE, NY.

The Sphinx had a decidedly mixed bag of genes. Up in the front she looked like a woman, but below that was a lion's body, a serpent's tail, and an eagle's wings. She earned her occasional huge meal by asking each passerby a riddling question. Prior to the arrival of Oedipus, each one had failed to come up with the correct answer. As a result, all were promptly throttled and devoured. Her question was: "What creature has only one voice, walks sometimes on four legs, sometimes on three, and sometimes on two, and is weakest when it walks on four?"[3] When wised-up Oedipus answered "Man," the Sphinx took that correct answer of his poorly, to say the least. She jumped off the cliff and — eagle's wings still tucked in —fell hungry to her death in the valley below.

Ingres was a better painter than he was a teacher. His students' works were often merely conventional and lacking in evidence of his ability to bring paintings to life. There was another reason for the failure of Ingres' followers to do well. The Classical impulse in painting was no longer as suitable as it had been to the changing social world of their time. There was the beginning of a greater valuing of "heart" than of "head," more interest in spirit and soul than in mythical or real factuality.

In the art-world fight between Classicism and Romanticism, Ingres represented the established classicists. The big name among the pushy newcomer Romantic painters was Delacroix. "Ingres pursued the fight of classicism against romanticism, of line against color which was to be for many years the chief topic wherever artists met.... Artists, collectors, and critics were openly divided into two hostile camps, between which the followers of both painters did their best to widen the breach."[4]

Turn now to a picture that still moves us, Eugène Delacroix's *Christ in the Garden of Olives*, 1827 (Figure 5-3). It shows Jesus with his eyes cast down and a hand raised up in a gesture to three sorrowful sad angels. In his time, the picture got mixed reviews. "One writer found an 'ineffable poetry' derived of 'genuine inspiration' in the three angels who have come to announce to their divine master His unhappy destiny.... [However, the eminent critic] Delacluze concluded: 'He has made a Christ in the Garden of Olives in which one [finds] some angels as gracious as little English schoolgirls'.... For many of Delacroix's contemporaries ... even a modest deviation from neoclassical orthodoxy was unacceptable...."[5]

The painting is realistic in the sense that he portrays the anointed Jesus and the unhappy angels as though they were painted from life or from photographs. Photography was not yet seen as a threat to painting. It was a tool that allowed models' poses to be captured and remain available whenever the painter chose to do his work. The author of *The Attainment of Delacroix* comments: "The essential purpose of Delacroix's *boules*, his work from photographs, his belief in memory drawing, or any other procedure he adopted was to capture what he called the 'tangible.' His ultimate aim, however, was not to achieve literal description but to unite the two worlds that had equal claims to his loyalty, the one rational and objective, the other subjective and intuitive.... The final product was to be at once a specific object and a stimulus to the imagination...."[6]

Those two worlds—the *rational and objective* and the *intuitive and subjective*—were still more or less united in the early nineteenth century of Delacroix. Although there was a crack between them, it was not yet the division that started later in the nineteenth century and broadened in the twentieth. "Delacroix never claimed the label 'romantic,' which he judged too reductive.... To his contemporaries, he was ... in the avant-garde [of Romanticism].... The tail of the [Classical] school was then dragging its

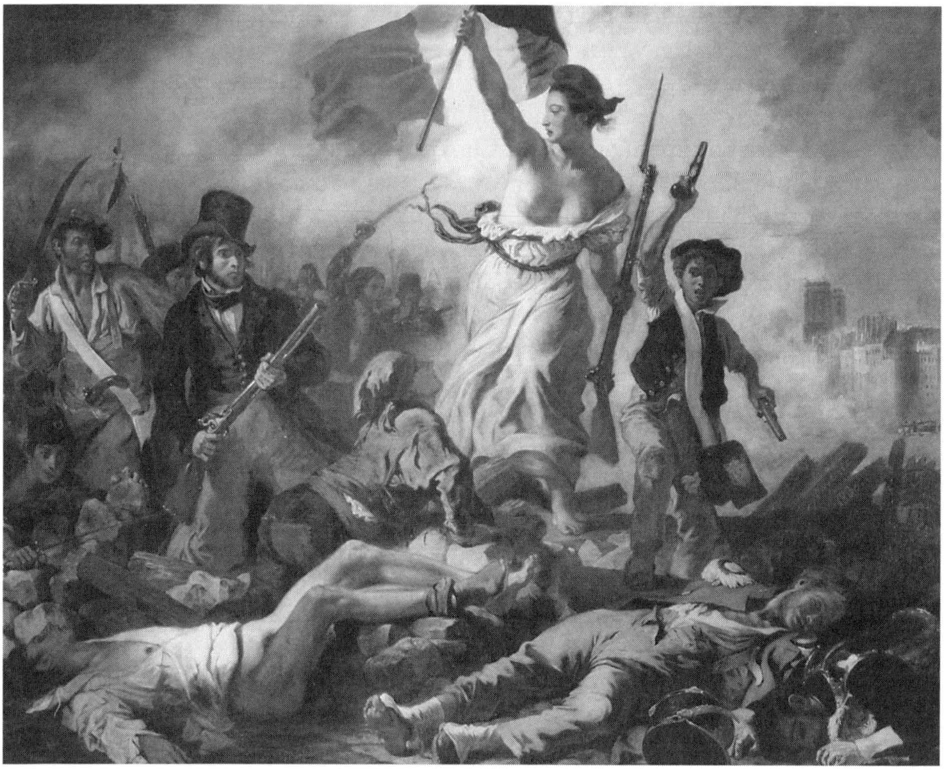

FIGURE 5-4. **Eugène Delacroix (1798–1865). *Liberty Leading the People*, 1830. Oil on canvas. 260 × 325 cm. Louvre, Paris, France.** PHOTOGRAPH COURTESY OF ERICH LESSING/ ART RESOURCE, NY.

last coils in the academic dust, and its paintings were nothing but feeble copies of Greek or Roman [works].... [In response to some of his works, the classicists] let out cries like an owl in sunlight, and the most comical outbursts had free rein: art was lost, there were no more sound traditions, the barbarians were at the gates."[7]

Painters like Delacroix may have used the word "objective" to refer to art, but they were not likely to use it to allude to science and technology. In fact, this artist was one of the many who had no interest in that subset of objective knowledge. It turns out that this disinterest cost Delacroix dearly. Some of his paintings fell apart — almost before his eyes. "Despite the deterioration that his paintings underwent within a very short time ... he could never resolve to abandon the disastrous use of bitumen, or of thick oil that was convenient for him or to refrain from going back over fresh spots that he wanted to change. The simplest elements of physics, of chemistry, of mathematics in the matter of perspective were unknown to Delacroix, for whom the exact sciences were a closed book."[8]

His *Liberty Leading the People*, 1830 (Figure 5-4), is certainly best described as a Romantic picture. The concept of Liberty is personified as a vigorous young woman. She strides forward, one hand rallying her followers by raising her fluttering flag, fixed bayonet on the musket in her other hand. As men and boys follow her, they pay no heed to the fallen dead and injured in front of them.

As usual, many critics thought it their necessary burden to be derogatory and disparaging. "[Most] ranged from neutrality to hostility. The figure of Liberty was condemned as clumsy and ignoble. The dead man at the left was dismissed as 'killed eight days before the conflict.' And then there were those who found [it] subversive in its 'proletarian' emphasis."[9]

Note that critics had no complaints about one feature of the picture. Wasn't it odd (and thus calling for reviewers' comments) that Liberty's breasts were exposed? After all, women then, as now, did not normally reveal their bosoms in public. No, it wasn't strange. That's because Liberty was not a contemporary woman of that time but a mythic person about whom Delacroix was creating a fable. As with portraits of Greek goddesses, her exposed body represented an abstract Romantic view of women's attractiveness in the eyes of men, rather than one that was titillating because it represented something real. Only a few decades after Liberty's appearance, the spirit of the times had changed enough for daring painters to portray real young women without clothing. Such pictures were seen by scandalized critics as portraits of painted ladies, trollops and prostitutes. Indeed, such works were not only artistic but also the precursors of what has now become Pandora's open box of mass-produced erotica and pornography.

At the 1855 Paris World's Fair, viewers "favored the classicism of Ingres and his numerous followers ... but seemed only reluctantly aware of Delacroix's greatness.... [In contrast, Jean-Franscois Millet's rustic paintings] showed a sentimental interest in rural life that was intolerable to people of so-called taste and refinement. 'This,' declared Count Nieuwerkerke, the Imperial Superintendent of Fine Arts, 'is the painting of democrats, of those who don't change their linen, who want to put themselves over on men of the world; this art displeases and disgusts me.'"[10]

A rather different connection between democracy and loathing can be learned from *The Anatomy of Disgust*: Democracy is "based less on mutual respect for persons than on ... certain styles of contempt.... The high begin to suspect anxiously that they are often seen as contemptible ... by the low. The high fall back on disgust for the low.... Once the low rattle their chains or are granted political equality, the high's complacent contempt gives way to a disgust prompted by a horror of the low."[11]

Count Nieuwerkerke, poor fellow, had to live in changing times that forced him to contend with the work of painters wearing what was probably dirty underwear. Had he lived a hundred years later, his aversion would have become utter revulsion, as there was much more democracy, and there were artists who made art from human waste products.

# PART III

# THE CHANGES BEGIN: THE SECOND HALF OF THE NINETEENTH CENTURY

### *Chronology of Technological and Scientific Developments*

1859: Darwin's *On the Origin of Species* published

1857: Pasteur proves fermentation is caused by living organisms

1865: Lister initiates antiseptic surgery

1866: Nobel invents dynamite

1869: Galton writes treatise on eugenics

1873: Color photographs developed

1876: Bell invents telephone

1877: Edison originates phonograph

1883: Koch describes preventive inoculation against anthrax

1892: Diesel patents internal combustion engine

1893: Benz constructs four-wheel car

1895: Roentgen discovers x-rays

## *Chapter 6*

# The Subject Beneath the Surface

> The expanding [nineteenth-century] scientific enterprise ... drew the support of a widening circle by its [practical results, which were] clearest, most dramatic, and most effective in medicine.... From the mid-nineteenth century onward, the conviction grew that the way to health passed through the medical laboratory.... By the end of the century the association of hospitals, universities, and laboratories [was the widely visible expression.] [Common to many claims] of the preeminence of science was the assumption that every deficiency in man's world was definable as a problem to which the correct ways of knowing would supply an appropriate solution. The staggering optimism of this article of faith endured into our own times.... Ultimately it promised that the organized use of intelligence, through its procedures, would perfect man.[1]

That kind of overreaching is called "scientism."

However, although many people late in the nineteenth century "embraced science warmly, fears were voiced from several quarters. The increased doubt which accompanied scientific education brought 'eternal values' into question.... [In] his Confessions in 1888, [George Moore] mourned.... 'The world is dying of machinery; that is the great disease, that is the plague which will sweep away and destroy civilization.... One can even say, but on pain of being stoned, that science has become the original sin of civilized man....'"[2]

It is hardly surprising that a changing world showed itself with big shifts in art. Painting had been "seen as a processing of nature, a re-creation and re-construction of nature. [However, the unrest of the late nineteenth century] was rooted in a deep disturbance of man's innermost attitude toward reality. Nature had ceased to be familiar, self-evident.... Art could no longer confine itself to reproducing nature, but had to define and represent the new reality, to make it visible."[3] "Modern art began with ... the realization that the visible is merely the surface of reality, that the outside world can be defined only with reference to man's inner world, and that this inner world is as tangibly real as the other...."[4] In 1863, an art critic expressed the changes like this: "The object of painting is to express ... the society which produced it.... As a result, painting is not at all an abstract conception.... It is a fragment of the mirror in which the generations each look at themselves in turn, and as such it must follow society."[5]

The aspects of the late-nineteenth-century world that changed and induced art to do the same — those aspects were mainly the way that people lived in cities. "[Modernity was] typically the experience of urban life in the new modernized metropolis— Paris, London, Berlin or New York. Paris in particular underwent a massive rebuilding program, involving new urban institutions and technologies such as the department store, the public hotel, the hired cab, street lighting and drainage, and major new public spaces such as the boulevard and the public park."[6] Also, tourism "through a countryside newly accessible by train [came into being, as did] the mass exodus from cities during summer months...."[7]

Although the case will be made here that much of Impressionist art was an upbeat view of the exciting, changing world, a rather different view was dominant in the twentieth century. Impressionism was thought to be "an emotionally impassive art of 'optical realism....' [It was concerned with] being objectively and even scientifically true to visual reality...."[8] Such a view of Impressionism was another case of mistaking the messengers for the message. The message was indeed that overwhelming changes resulted from the new science and technology. Impressionists responded by portraying what was stimulating about the new world. That hardly made their work an emotionally impassive art of optical realism.

However, some who followed the Impressionists did turn their emphasis away from the outer world and moved closer to the subjective life of feeling. In 1886, a writer/critic characterized those who came after the Impressionists like this: "The essential aim of our art is to objectify the subjective (the externalization of the Idea) instead of subjectifying the objective (nature seen through the eye of a temperament)."[9]

In the chapters that follow, we will look first at Edouard Manet, who painted a new, more open view of things that had once been hidden. Then we will focus on a pair of the earliest hasty pictures by Claude Monet and Auguste Renoir. Next we'll turn to Berthe Morisot, some of whose paintings represent the very feeling of rapid change. After that, some of Edgar Degas' studio-made pictures show us women as other than sex objects, women who actually have lives of their own. Following that, we turn to Post-Impressionists Vincent van Gogh and Paul Gauguin, who emphasize the more inward view near the century's end. Finally, we will look at some of Paul Cézanne's turn-of-the-century paintings of an eerie, remote, objective world.

## *Chapter 7*

# A New Realism: Edouard Manet

Cast your eyes upon Edouard Manet's *The Old Musician*, 1862 (Figure 7–1). To be sure, art critics had problems with this picture when it appeared. "When compared to paintings on a similar subject, the bohemian allegory or gypsy group, [the painting] appears incompetent: the tree is cut-off, the figures abruptly organized, the characters self-absorbed or in blank day-dreams."[1]

The art community was still uncomfortable with the picture two decades later. "[One critic wrote,] 'Manet is merely a painter ... of fragments—devoid of ideas, imagination, emotion, poetry or powers of draughtsmanship. He is incapable of composing a picture....' [Others saw the painting] as marking an abrupt break with the art of the immediate past and with the kinds of expectations that the public had of painting.... To see *The Old Musician* this way is to echo the view of art..., in [Max] Weber's words as 'an inner-worldly deliverance from every-day life....' By the contemporary norms of the Salon, the painting is technically crude, but this is *related* to a concern to represent the contradictory experience of *modernité*."[2]

Apart from presumed technical problems, the painting clearly reflects the idea that the world was changing. That can be seen in the way the central figure (who carries both the name and the weight of the picture) looks out from the picture into the viewer's eyes. Although the other people look at each other or at him, he leaves his surroundings to look out at us with an expression of alienated perplexity, as if to say: "Yes, I am here. Yes, it is strange. No, I can't understand it."

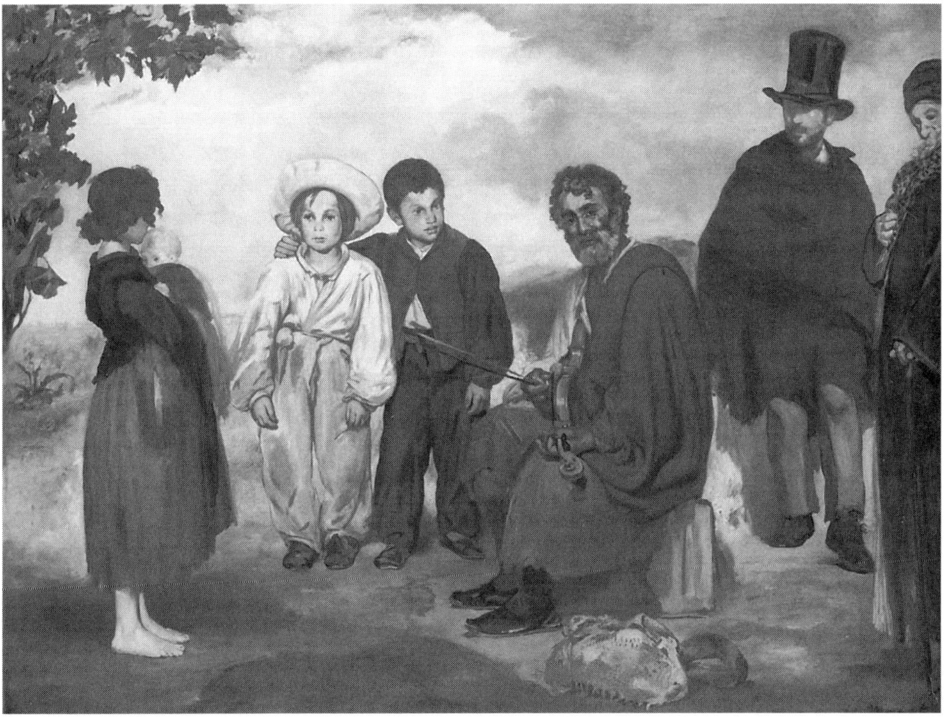

FIGURE 7.1. **Edouard Manet (French, 1832–1883). *The Old Musician*, 1862. Oil on canvas.** CHESTER DALE COLLECTION. IMAGE ©2004 BOARD OF TRUSTEES, NATIONAL GALLERY OF ART, WASHINGTON.

The old musician's puzzled alienation is emblematic of the *changes* in the world that were increasingly evident to people in the middle of the nineteenth century. "As inventions speeded travel and the mails, identified diseases and modernized production, the shock of the new became the signature of the age.... The population explosion radically altered the dimensions of the market ... millions migrated from country to city...."[3]

"For a substantial part of the period 1850–70, life in Paris was overshadowed by ... the systematic demolition of narrow, rabbit-warren-like streets, and their replacement by eighty-five miles of straight, wide, tree-lined avenues and boulevards.... On a lesser scale, similar reconstruction took place in Lyons, Marseilles and Le Havre; nothing like this had been seen since Roman times."[4]

However, "refuse still had to be removed by the traditional night-time army of collectors, buried in cess pits, or illegally dumped in streets or gardens."[5] For us in the twenty-first century, the march of progress has brought daytime trash collectors who drive Dumpster-emptying machines. Ever since it started accelerating in the nineteenth century, progress has been a mixed bag—one that contained treasures, terrors, and trash.

Look now at Manet's *Déjeuner sur l'herbe* (Luncheon on the Grass), 1863 (Figure 7–2). In *The Birth of Modern Painting*, we read: "What strikes us in Manet ... is a kind of total presence of the image.... The scene is played here and now, before us ... it is the fact of the moment, not a history but a presence ... the woman in [the center of the picture] does not look at her companions but at the 'flash' which take her likeness.... It was ... the end of models which the individual does not meet in real life but which he can reconstruct.... [It was the end of] legends and myths, great actions and historical figures, ideal representation of beauty, ideal objects of desire."[6]

However, *Déjeuner sur l'herbe* was also "considered an affront to morality and to art in its adoption of old-master imagery for a scene evidently

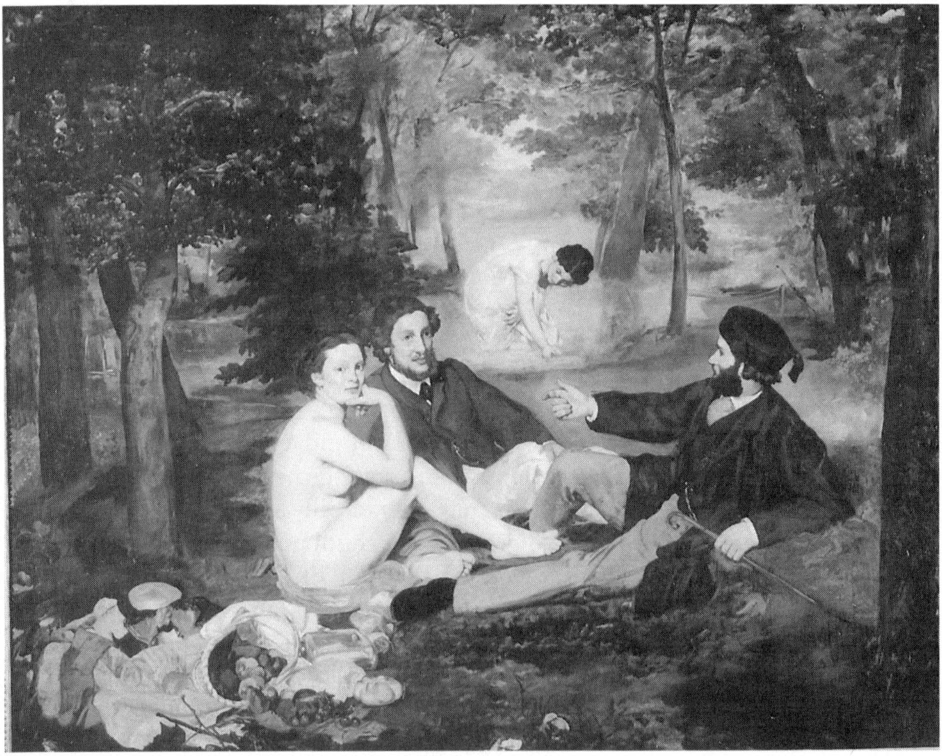

FIGURE 7.2. Edouard Manet (1832–1883). ***Déjeuner sur l'herbe*** **(Luncheon on the Grass), 1863. Musée d'Orsay, Paris, France.** PHOTOGRAPH COURTESY OF ERICH LESSING/ART RESOURCE, NY.

modern and also [an affront] in its broad, flat, graceless presentation of it. The sacred, if pagan, here looked merely bohemian.... Manet had based his picture on Titian's *Venus of Urbino* yet seemed to caricature it stylistically ... and [to mock it] morally by showing a very real model, probably a prostitute, in a challenging pose and mood."[7]

The nakedness of the woman is the first thing that catches the eye of most viewers. She sits calmly beside two well-dressed men who pay her no heed. What on earth are these people doing? To all appearances, the four (a second woman is in the background) are simply having a picnic lunch. There's no hint that the nudity implies an after-lunch roll in the hay, so that may be why the picture was less of an affront than the pictured prostitute in Manet's *Olympia*, 1863–65 (Figure 7–3). Nor would the fuss about *Déjeuner sur l'herbe* have had much to do with class. Well-dressed men would hardly expect youthful women of their own leisure class—young *ladies*—to take their clothes off during a picnic. So, almost by definition, the women must be lower class (although being bare is just as odd for them). But a painting of an outdoor lunch with naked women was still decidedly surprising.

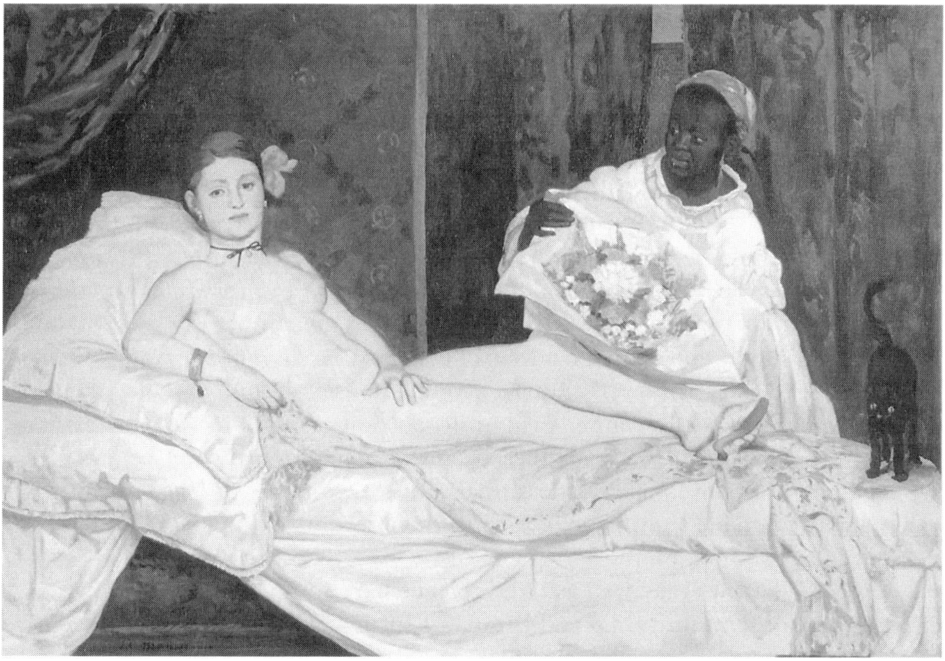

FIGURE 7.3. **Edouard (1832–1833). *Olympia*, 1863–65. Oil on canvas. Musée d'Orsay, Paris, France.** PHOTOGRAPH COURTESY OF ERICH LESSING/ART RESOURCE, NY.

The Enlightenment had changed some things but hardly everything. That men would gratify their prurient interest in lower-class women as sex objects was scarcely scandalous; it was common knowledge. After all, it was working-class young women who became trollops, tarts, harlots. The arrival of the twenty-first century has not changed this in the urban areas of non–Western developing countries. There, many women and adolescent girls sell sexual services to meet basic needs. (However, we should also note that in at least one developed country, Japan, some adolescent girls now sell sexual gratification for money they use to buy upscale merchandise.)

Viewers would have been uneasy about *Olympia* because men's interest in young female flesh was expected to remain *covert*. So the nudity in it, as well as that in *Déjeuner sur l'herbe*, would have provoked more a show of shock than the real thing. What viewers might have been eager to say to each other was, "I'm shocked. Shocked, do you hear!" The real shock was the calm honesty of Manet's pictures of naked real women, the openness to the objective truth that men were interested in young females as sex objects.

Look now at what scandalized Paris in 1865, the public appearance of his *Olympia*. Critic Kenneth Clark reminds us that Manet's predecessors normally "represented the body as existing solely in twilit groves or marble swimming baths."[8] *Olympia* was definitely *not* the image of an ancient unclothed goddess. Manet showed his model as a real-life harlot, an unclad "working woman" who stares calmly at the male viewer and knows why he is staring at her. This was a new kind of realism.

Shocked or not, people in cities like Paris were well aware in the last half of the nineteenth century of "the massive spread of prostitution.... During no period, before or since, was the sight of prostitutes so common.... Their presence [was] a logical result of the normal workings of the law of supply and demand.... Men of the middle classes had 'elevated' their wives to the position of ... quasi-virginal household nuns.... They therefore tended to slip rather easily into encounters with ... women of the working class.... [Many of] these women, forced to choose [either] deadening hours of ... labor at less than subsistence wages or [the evidently] more tolerable alternative of prostitution, ... came to populate the streets of the major cities."[9]

What, we may well ask, did science have to say about the changing world of femininity? Lo and behold, it turns out that science provided a handy confirming belief "that women were not fully equipped ... with the superior mental functions [of men], especially abstract thought.... The origin of woman's supposed mental weakness was to be found in the structure of her brain which was described as 'less wrinkled, its convolutions ... less beautiful, less ample' than the masculine brain.... Women were thought ... to stop developing intellectually at the onset of puberty when their constitutions became taken over by their reproductive functions."[10] Whoa! The young woman who was the model for *Déjeuner sur l'herbe* and *Olympia* had not only less wrinkled skin than did most of her male oglers, but she had a less-wrinkled brain? Science said that women were playing with a deck less full than that of men?

Science did indeed do that. Even "Darwin, in *The Descent of Man*, unhesitatingly accepted the notion of the natural inferiority of woman.... [He] had already expressed the opinion that ... the intellectual distance between men and women was bound to increase as a factor of natural selection."[11]

Perhaps we should be shocked — shocked — that science, like all other human endeavors, is subject to mistakes and deficiency. However, imperfect science (with its less than totally valid knowledge) *accepts* the possibility of error as an inevitable part of the search for truth. That very distrust (which led to the correction of found error) was central to the dramatic beginnings, in Manet's time, of the changes in the way people live in our time.

"Looking at [Manet's 1863] *Olympia* now, it may seem unlikely that this painting could have proved difficult or impenetrable for its contemporary public.... [More interesting is that critics'] discomfort with the supposed vulgarity of the woman depicted became embroiled ... with the supposed inadequacy of Manet's artistic technique.... The critic Théophile Gautier, for example, wrote: 'The color of the flesh is dirty, the modeling non-existent. The shadows are indicated by more or less large smears of blacking....' [So there is a strange] slippage at work, which conflates the character of the picture with the supposed character of its subject alone."[12]

The criticism by Gautier and others was hardly the first time that the messenger (in this case, Manet) was blamed for the message (in this case, that attractive young women who appear to be sexually receptive are always of interest to men). Nor were professional critics the only source of complaints about this messenger. Listen to these words from an established great painter looking down his nose at the newcomer: Delacroix said that "[earlier painters' brushstrokes] 'had nothing to do with the monstrous abuse of brushwork and the slack manner of the painters of artistic decadence.... Monsieur Manet has all the requisite qualities to be rejected unanimously by all the juries of the world. His keen sense of colouring stabs the eyes like a steel saw; his figures are cut out as if by a die, with a crudeness unsoftened by compromise.' [Likewise, the great]

Ingres never ceased to complain of a Paris 'delivered up to the most violent artistic anarchy.'"[13]

What Delacroix then called a "monstrous abuse of brushwork" is now viewed as brushwork that was deliberately painted to be visible. That was one of the "technical" innovations to be seen in Manet's pictures. Another was the diminution of depth in the scenes portrayed. As Michael Crichton tells it, among "dissident figures in this history of artistic fidelity of nature ... the first of importance is Edouard Manet, the well-to-do Parisian dandy, the artist famous for his waistcoats, his delicate women, his luminous light.... Manet was a precursor of Impressionism, a major movement to which he stood in uneasy relationship — [much as Jasper] Johns does to Pop Art. Manet, like Johns, was notable for his willingness to let a painting be flat, lacking the illusion of depth."[14]

Critic Peter Schjeldahl has some pertinent comments about *Olympia* and Manet. "When one really looks at his paintings, ideas vanish. Beauty, poignancy, and a sort of happy shock prevail.... *Olympia is wearing six things,* a high-heeled slipper..., a ribbon choker with a pearl attached, a pair of earrings, a bracelet with another dangling jewel, and a flower ... in her pulled-back hair. Every item renders her more naked...."[15]

These words of Schjeldahl's certainly sound right, don't they? When we look at the woman shown in *Olympia*, ideas simply vanish and we experience her erotic nakedness. But we are only a century and a half away from her. And — in that young women, now as then, normally do not let themselves be seen unclothed in public — we live in the same world as she did. However, the truth is that, then or now, near-bare equatorial tribesmen would not take particular note of her nakedness. Nor would they find her nudity to be erotic. So it is closer to the truth to say that ideas related to pictures like *Olympia* do not really vanish. Rather, we project such notions into the painting. They become *embodied* within the picture, from which we introject their erotic quality back into our own minds. It is the viewer's deeply embedded belief that Olympia is *sexually receptive* that makes her not just unclothed but utterly naked.

## *Chapter 8*

# Mad Haste with Impressionism: Monet and Renoir

Both Claude Monet and Auguste Renoir painted an early picture of the same small outdoor recreation place. Those two works provide us with a way of seeing something new about the motivations behind Impressionism. An oddly arresting feature in the way both pictures were created, the use of attention-demanding brushstrokes, will give us that access. We know that the Impressionists themselves were interested in the properties of outdoor light. "The typical Impressionist ... was prepared to find a *motif* anywhere — in a haystack or a lily-pad — it did not matter because his primary interest was in the effects of light."[1] But painting the effects of outdoor light, which would soon change, required *haste* to capture them. We will see that haste to capture reality before it changed was more than a necessary method. It was also a metaphor for something broader. It was a way of portraying the struggle to deal with the rapidly changing world of the nineteenth century. That was another kind of significance, one that was hidden behind haste, embodied by this new way of painting.

These two painted "impressions" are of a place that provided "facilities for bathing and for boat hire [as well as] a floating restaurant and dance-hall and riverside tables for eating and drinking.... [These were] fashionable attractions for the city-dweller in pursuit of leisure.... The pictures that Monet and Renoir painted [there] are celebrated as making a decisive stage in the development of Impressionism."[2]

One of Monet's several paintings of this place is *La Grenouillère* (The Frog Place), 1869 (Figure 8–1). This one is only 29 inches by 39 inches. Renoir also made a number of small paintings there. One of his carries

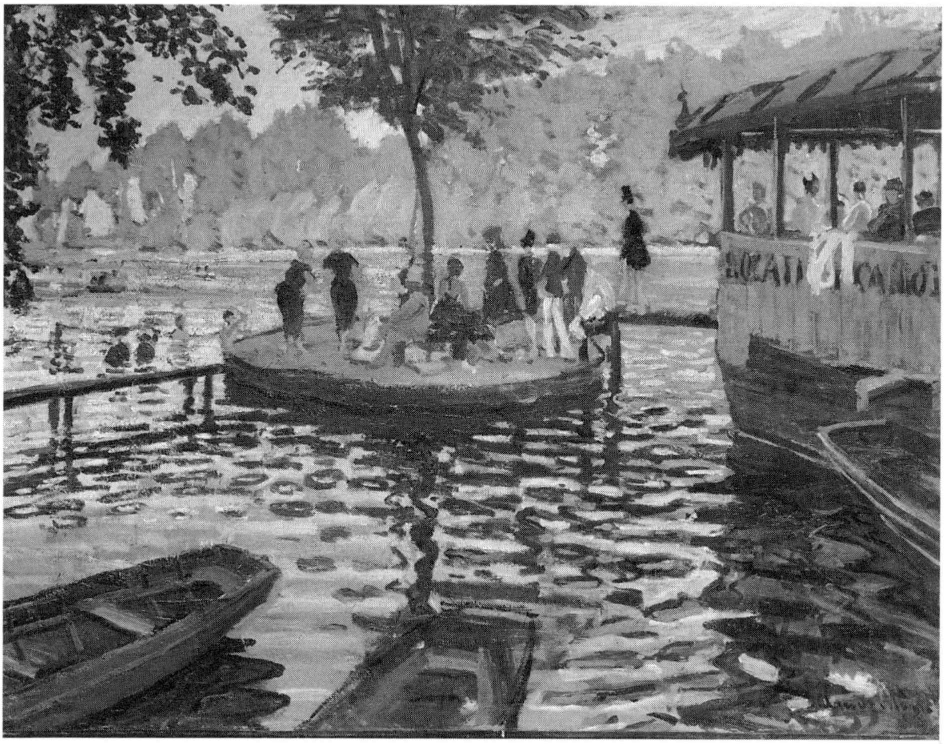

FIGURE 8.1. Claude Monet (1840–1926). *La Grenouillère*, 1869. Oil on canvas. 29⅜ × 39¼ in. (74.6 × 99.7 cm.)  H.O. HAVEMEYER COLLECTION, BEQUEST OF MRS. H.O. HAVEMEYER, 1929. (29.100.112)

the title *At La Grenouillère* (At the Frog Place), 1869 (Figure 8–2). We will see that the modest size of these pictures may have been an indirect contributor to the style called Impressionism that more or less began with them. At this place "a world of people from different social backgrounds came together with the sole aim of having fun, dancing and boating."[3] The unromantic name of this place refers not only to the amphibians who lived there around the clock, but also to the working-class young women visitors, women who might be sexually receptive to the men with more money who went there to hit on them.[4]

Art historian Meyer Schapiro tells us that the distinctness of Impressionist painting is due, in part, to the kind of brushstrokes they used. "The strokes seen together fabricate a surface web — a film or crust of pigment that here, more than in earlier art, exists as distinct in texture from the image-object.... The Impressionist produced two modes of appearance.

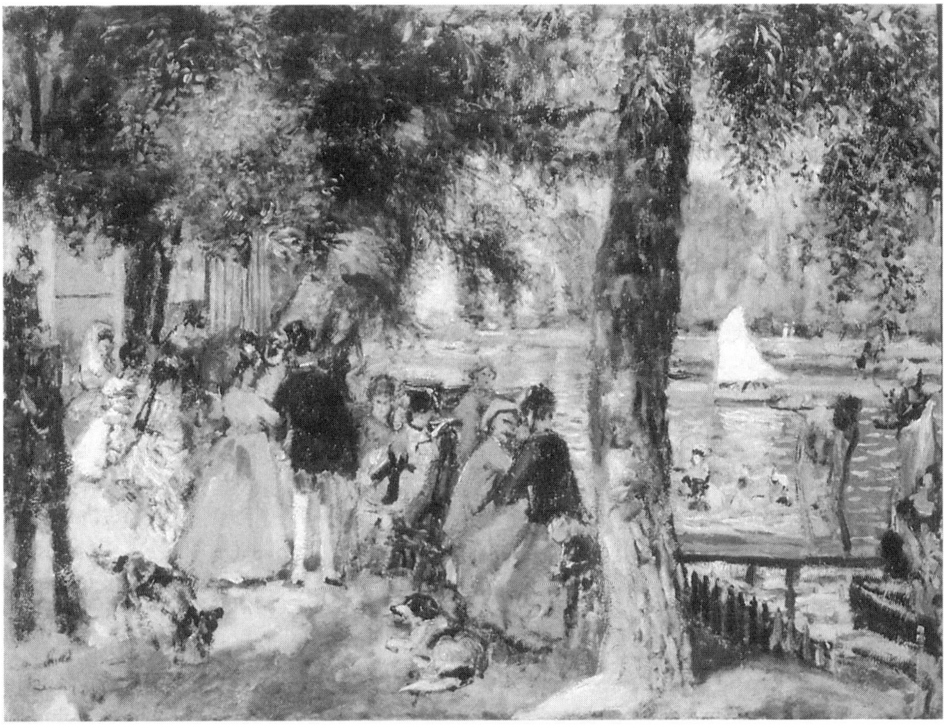

FIGURE 8.2. **Auguste Renoir (1841–1919).** ***At La Grenouillère*****, 1869. Hermitage, St. Petersburg, Russia.** PHOTOGRAPH COURTESY OF SCALA/ART RESOURCE, NY.

First, there is the image aspect — the scene that you make out in the fictive space of the painting. But you are also made aware that the layer of paint is thick.... Before Impressionism, artists [usually created] a smooth, shiny surface.... The canvas became transparent, an invisible plane, like a mirror."[5]

The authors of a book about nineteenth-century French painting describe the use of such brushstrokes as somewhat of a quandary. "The particular distinctiveness of [works such as those Monet and Renoir made at La Grenouillère] lies in the relationship between the brightness of the colour and the breadth of the tonal range they display on the one hand, and on the other the apparently rapid, 'broken' brushwork with which they are executed.... It is by no means clear how we as viewers might distinguish between our response to the animation of the *scene* represented and our response to the animation of the representing *surface*.... In Monet's painting ... the figures ... read less as individuals than as tokens— nor do they ever quite lose their identity as groupings of brushstrokes."[6]

It is indeed the peculiar large size and vivid coarseness of some of the brushstrokes in these early Impressionist pictures— a distinctness that calls our attention to the surfaces of the paintings— that needs an explanation. Brushstrokes started to show up, to show themselves literally, in pictures by Edouard Manet only a half dozen years before these. However, in claiming the viewer's attention, the brushstrokes in these pictures of La Grenouillère are much more insistent than are Manet's.

Look, for example — in Monet's version of the scene — at the white trousers on the man at the right end of the central round island. Look also at the figure in white to the right of that man. Although the reproduction here is so small that you may have to peer closely to see what seem to be people in white, the somewhat larger original painting would require no staring to bring your attention from the scene portrayed to the coarse and incomplete white brushstrokes on the surface of the canvas. Attention-grabbing paint-streaks can also be seen in Renoir's painting, as well as elsewhere in Monet's.

The method used by outdoor (Impressionist) painters provides an obvious reason for such a "rough" portrayal of things like the white clothing. They simply had to apply paint in mad haste before the movement of the sun's light changed everything they were viewing. Note, however, that there are other parts of the pictures— in Monet's, the pink hands and feet of some of the people — that are painted with finer detail. So the idea of mad haste that demanded the broken brushstrokes may be something more than an immediate explanation. After all, the artists— at their later leisure —could have "polished" the details of pictures like these, couldn't they? They didn't do that; doing so just didn't seem "right." As we shall see, mad haste to put roughly right paint on canvas is not only a surface explanation and a seeming excuse, it is part of a deeper account for the new painting method called Impressionism.

Also pertinent to this puzzle is the following. There is reason to believe that Monet and Renoir planned to make physically larger pictures based on these small ones, but failed to follow through. One of the reasons they didn't carry out such plans may have been that larger versions of the seemingly hasty brushstrokes would simply not have "worked" in the sense of being credible to the pictures' viewers. So it is likely that the artists, in some sense, "intended" the brushstrokes to be hasty.

Before examining a reason (beyond mere haste) for the hard-to-overlook "roughness" of these paintings, let's look at the *idea* of the increasing prominence of brushstrokes in the painting of pictures. What, exactly, is the big deal about emphasizing the fact that paintings have surfaces? After all, none of the viewers is an outlander who just arrived via the boat from Mars. We all know how babies are made, and we all know that pictures are made by putting paint on the surface of canvas.

Nonetheless, the issue of the relationship of two-dimensional paint and pictured three-dimensional objects is one that grew for a century. It more or less started with Manet's newly visible brushstrokes (in the middle of the nineteenth century). As we will see, it grew to the (mid-twentieth-century) creations of Jasper Johns that include on their surfaces both painted, flat, two-dimensional objects such as targets, and real three-dimensional objects such as sticks. It seems likely that in every case this conflict between picture content and surface served a purpose. It was a way of representing some *outside significance*— some meaning not shown *within* the picture itself—that was of import in the painters' society at the time.

What we have here is not a case for narrow-minded emulation of Sherlock Holmes. Any armchair-detective must resist the temptation to examine the crooked growing timber of our knowledge about art, and then focus only on the wood beneath, or only on the tree's surface bark. Such a chair-sitter should not simply insist either that "There is just too much of such surface, not enough of the real timber that is deep inside" or "Aha! The bark is not quite missing, but there's too little of it. Never mind the tree, the bark on the surface is what we see."

Something like the latter was done near the middle of the twentieth century. Then, many in the art community acted as though it was only the surface of a painting that was all-important, and it was somehow wrong to "fake out" the viewer by portraying seemingly three-dimensional things. (When we get to a few of the Abstract Expressionists— Mark Rothko, Lee Krasner, and Jackson Pollock — we'll peek beneath the surface of some of their flat pictures to make more sense out of them.)

To make sense of the *conflict*, in the Grenouillère pictures, conflict between the surface of the paintings and the solid-seeming objects within the pictures, the following should help. Drawing attention to the surface of Impressionist paintings was a way of reinforcing the basic idea that was itself behind Impressionism: The new style of painting was a way to represent both interest in and uncertainty about all the rapid transformations that were making the world *modern*. It illustrated the view that "All this change is exciting, but it is also troubling."

We know that *ideas* are difficult to portray explicitly in a picture (unless the title words provide help). In themselves, concepts are too abstract. How can you paint a (static) picture of *rapid change*? On the other hand, if these painters had lived in a world that was changing so little that it seemed eternal and immutable, it would not have been of any interest to attempt to represent change in their pictures. For those who live in a stable world, familiarity and routine serve quite adequately without raising the idea of significant change. (We say, "Are you new at this job or something? That's just the way we do things here!")

For the actual decades preceding the 1869 of these pictures, things were changing so rapidly that the painters' world may have seemed ephemeral and full of contingency. Both the fact and the meaning of all that change would have been of major concern and interest to many of the people in the world of these painters.

Impressionism is normally regarded as painting in the open air "to capture the visual experience.... A scene was chosen but normally left without ... compositional rearranging or completing...."[7] What is suggested in this chapter is that (no doubt without explicit intent) what the Impressionists found was a new way of representing the new interest in their rapidly changing world. Overdramatized brushstrokes demand a share of the viewer's attention, even though they have little meaning in themselves. They make a picture so sketchy and crude that the legibility of the portrayed images is almost threatened. Yet they are ultimately seen to be parts of meaningful portrayed objects, such as people standing on a pleasure-island in these pictures by Monet and Renoir. Thus it was more pertinent than it was paradoxical for Impressionist painters to use brushstrokes of questionable significance in themselves. What such strokes do is raise the idea itself of the painter's urgent haste to capture a scene before it changes. They bring the idea of change closer to the foreground of the viewer's attention, there to compete with or augment (by adding the excitement of uncertainty) the objects portrayed.

It needs no saying that these pictures — of this tiny leisure world that was a tiny precursor to places like Coney Island and Disneyland — do not provide viewers with grand notions of "The Meaning of Life as It Changes." What they do show — in the animation within the three-dimensional scene portrayed — is something of what life at new locations like La Grenouillère was like. The people pictured were those with some newfound freedom and money to find new friends and enjoy life. Thus these people took a view of their nineteenth-century changing world that was far more upbeat than it was a downer.

The pictures' brushstrokes, however, demand a share of the viewer's attention — demand it for purposes that are not directly apparent. So they also raise some confusion in the minds of viewers, confusion that reflects the uncertainty that comes with rapid change in the way people live. That confusion and uncertainty about rapid change — which gets symbolized in Impressionist pictures with disconcerting brushstrokes—conveyed the very idea of the need for mad haste to capture the appearance of a scene before it ceased to be what it was. So the basic method of Impressionism itself—writ large—was at one with broken brushwork in conveying the feeling of the rapidly changing times in the last third of the nineteenth century.

In the decades after the Impressionists, science continued to change the world. It began to seem more a world of threatening new artifacts and empty consumption, a place where the joy of living had been lost. It was only then, as we will see, that many painters turned increasingly away from the changing world, and others began to accentuate its downer side.

# Chapter 9

# The Fragile Moment: Berthe Morisot

In *The End of Time: The Next Revolution in Physics*, the author writes about his belief that time itself is literally not *real*. Instead of time, nature has fashioned a huge set of *fixed* "three-dimensional snapshots,"[1] each of which he calls a "Now." Time itself is nothing but an illusion that we have from being part of a number of Nows—with nothing at all between them.

Not in that fantastic literal sense, but in a figurative way, we all use that idea. We suggest that photographs and paintings of the objective world stop time to give us the equivalent of a Now. Berthe Morisot's portraits, for example, stop time in that ordinary way. Other paintings of hers, however, do something more than just record the here and now. They give us a feeling that she has somehow captured a delicate transient event that is so unique that not only will it disappear in a moment, we will never see its like.

In her painting *Eugène Manet on the Isle of Wight*, 1875 (Figure 9–1), Morisot did just that. She apprehended her husband "looking out of an open window over a small garden to the harbor and the ladies passing by. It is a simple, everyday scene ... [with] the long rectangle formed between the top of the garden fence and the bottom of the open sash window, with its glimpse of boats and a little girl and a woman...."[2]

But the artist captured something more than that. She caught something revealing that does not occur every day, something that will vanish in a moment. She collared her newlywed husband staring at the face of a young woman strolling by. Neither viewers of the picture nor Morisot as

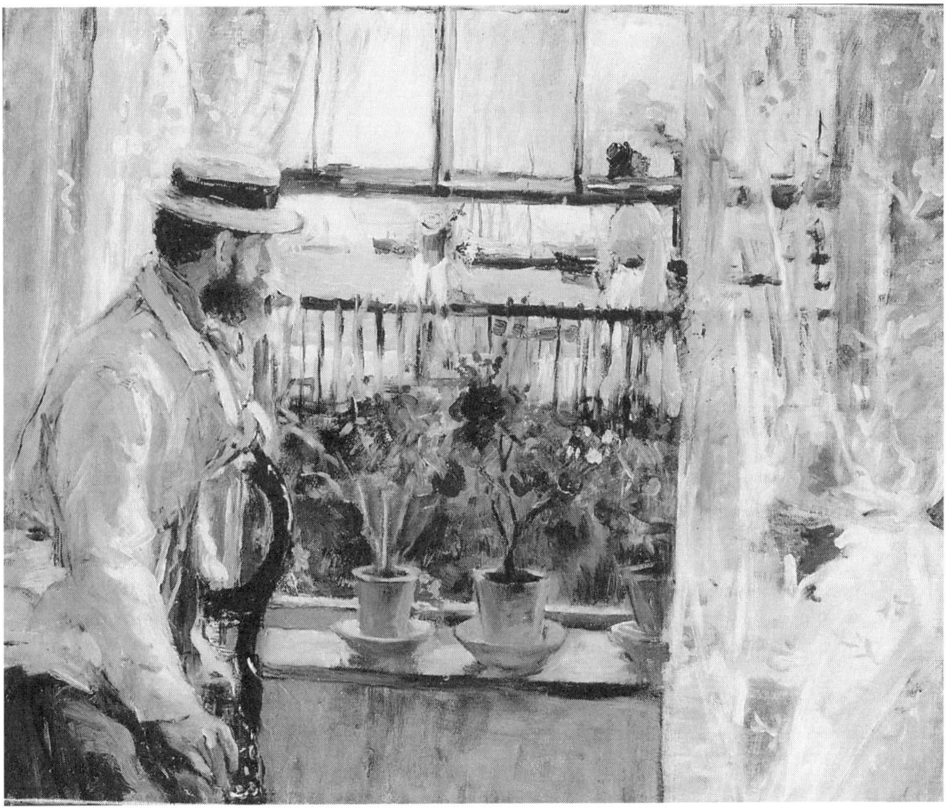

FIGURE 9.1. **Berthe Morisot (1841–95). *Eugène Manet on the Isle of Wight*, 1875. Oil on canvas. COURTESY MUSÉE MARMOTTAN, PARIS, FRANCE/GIRAUDON/BRIDGEMAN ART LIBRARY MMT160284.**

wife/painter can see the woman's face, but it is clear that her husband's eyes are captivated by the stroller's appearance.

Although we cannot know whether Morisot afterwards confronted her man about his roving eyes, that may be unlikely. At about that time, when she was in her middle thirties, she wrote to her brother: "I have found a gentle and kind husband, who, I believe, loves me dearly. At last, I feel that life may be really happy."[3] Married men — of the moneyed class that she and her husband belonged to— often did more than merely look at many young women as potential sex objects. They sought and bought the pleasure of intimately embracing them. Morisot might have faced that situation, or she may have looked the other way. Chances are that she simply knew that most men — whether or not they acted on it — have the wandering-eye syndrome, and found it best to ignore it after she painted it. What matters for us is that she seized a fragile moment that can be seen over a century later.

In *A Lady in a Park*, 1885 (Figure 9–2), Morisot has her way with time (and with us) without the help of strolling female passersby. This time it is her use of watercolors, or rather, the impression of hasty, broad watercolor brushstrokes that suggests that the artist was desperate to snatch what she saw before the lady sauntered off with her parasol. This time, the seemingly reckless swiftness of her brush is what has enabled the artist to enchant us.

Her *Femme et enfant au balcon* (On the Balcony), 1872, shows "a young woman in black ... with a red-haired child in a white dress beside her,

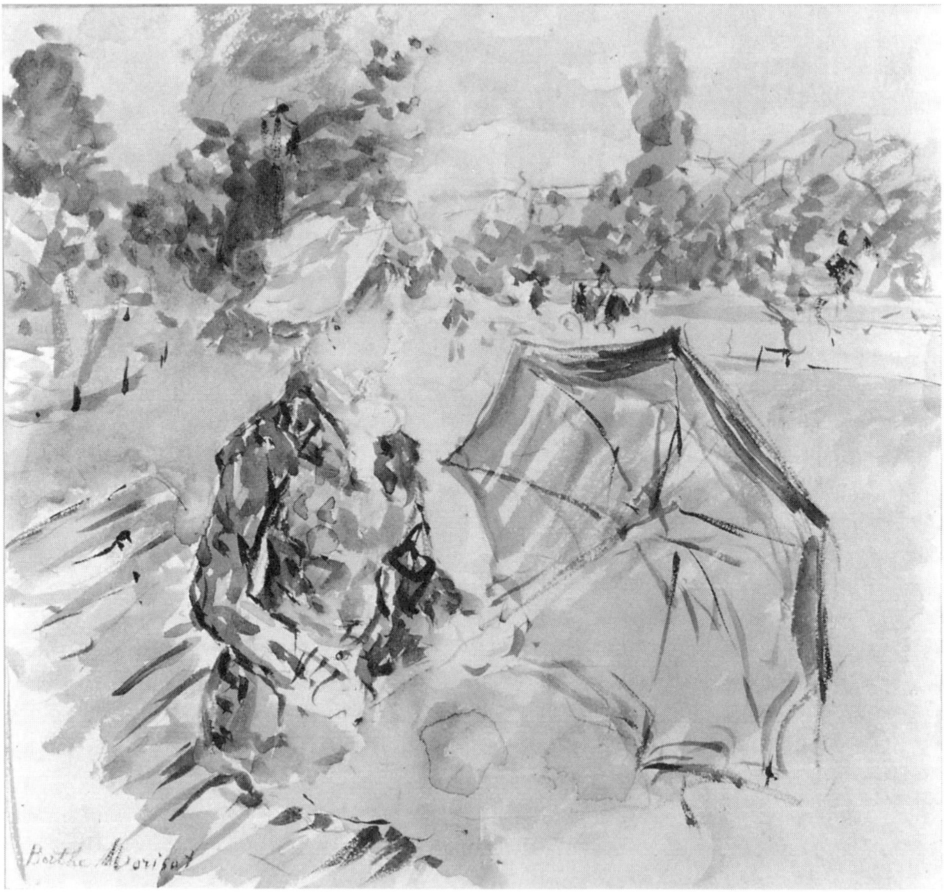

FIGURE 9.2. **Berthe Morisot (1841–1895). *A Lady in a Park*, 1865. Watercolor.** ALL RIGHTS RESERVED, THE METROPOLITAN MUSEUM OF ART, HARRIS BRISBANE DICK FUND, 1948. [48.10.8]

looking out over a garden that we cannot see.... The vast horizon of the city with the blue river glimpsed between the balcony and the empty Champ de Mars, give to the woman and child a strange dimension of mingled absence and presence, evoking the clarity of a dream."[4]

The picture has the clarity not of a dream but of a Now that we know will disappear for the rest of eternity. Yet it carries no suggestion that the Paris in the background had been changing and would continue to change. To the contrary, it looks like it will be there forever. Nonetheless, the painting captures what seems a fragile moment, a moment that won't last another moment. This time, it is the little red-haired girl (Morisot's niece) who performs the magic of representing change; we know she will skip away in an eyeblink.

Writing that "Euclid alone has looked on Beauty bare" (as did Edna St. Vincent Millay) is well-earned poetic license. What Euclid and countless others saw was *static* beauty, the beauty of mathematics, which seems timeless to us. In fact, some mathematicians think there somehow exists a real realm where mathematics and other truths did/do/will reside—reside without us—forever.

The beauty we see in some of Morisot's little-more-than-a-century-old paintings, and in some of the paintings of other Impressionists, is the very opposite of such static beauty. It is *vital* beauty that seems like it will be gone in an eyeblink. It is kin to what greets us briefly in the springtime when we are lucky enough to look upon otherwise bare-branched jacaranda trees just as their blossoms start to fall.

The transience of what is captured by Morisot and other Impressionists is one thing. However, to refer to their work as *beautiful* is something else. It is done here because the work of the Impressionists is so regarded by many people now, by more people than would use that kind of a word about the work of any other modern "school" of painters. "Beauty" certainly seems more of a subjective word than an objective one. Yet that people can and do largely agree—that many Impressionist paintings are lovely—suggests it has an objective nature as well. To explain beauty won't be attempted here, but we will examine why it is that most of us can agree about the presence or absence of such a quality in pictures from a given school of painting.

Let's inquire about the ways in which we are alike mentally—if indeed we are. Anthropologists refer to this as their "oldest and most vexing question: [Does] the psychic unity of humankind [exist]?... [Such a] commitment to psychological universalism [has been called] the cornerstone of 'The Enlightenment view of man.'"[5] Yet, as individuals, most of us know people whose minds—in terms of the *ideas* within them—are not at all like ours. And anthropologists know that the cultures they once studied (most of which have now disappeared) were similarly unlike one another. That suggests that claiming we are all alike — alike in the concepts we think with — is empty puff-pastry. What may indeed be alike for all of us are the kinds of categories— the containers for different types of concepts in our minds— such as those for objective ideas and subjective ones.

Nonetheless, most of the people that share a culture may indeed share a set of mental concepts. The external things that make up a culture — its artifacts, social institutions and directions for our interaction — are the embodiments of those shared ideas. To get back to the Impressionists, our present culture's ideas remain enough like theirs to allow us to understand their paintings. At the same time, our culture is different enough from theirs that we do not feel threatened by the rapid transformations that were central to their world. That combination —comprehension, and freedom from anxiety about change — plus their upbeat view of life is what spells enjoyment, and even beauty, for us in viewing Impressionist pictures.

Within the lifetimes of Morisot and her fellow Impressionists, there were huge changes in their world. Although it was rapid change over the decades of their lives, it may not have been visible over a given year, let alone month, day, hour, or moment. Fast or slow though it sometimes was, the people who lived then knew very well that the times were changing. As we saw in the last chapter, the haste itself of Impressionism's method symbolized feelings about change. Artists like Morisot also succeeded in *representing* this feeling of rapid change in their world by seeming to capture fragile moments before they disappeared.

We live in the most recent incarnation of the world of rapid change, change that results from the progress of science and technology. Pictures like these of Morisot's show something of past change without frightening us about that change. So they enable the love side of our love-hate relationship with change — they enchant us.

As we'll see in chapters to come, the hate side of change also looms large in our lives. Change that threatens the comfortable continuity of our own lives is not at all lovely. Change that threatens our values (let alone our living safely) is hateful when it is not frightening. Change that goes in one direction — and toward a better future — is something to which we give the name "progress." When change seems to go toward a worse future, the usual antonym, "regression," is not appropriate, as it suggests moving *backward*. The rapid changes that started not long before Morisot's time, and continue through our own time, are the result of the continuing expansion of objective knowledge that builds monotonically. We cannot characterize that expansion as either plainly good or just bad, as it has been both. More important, we can't know whether we will use it to make a better future or a worse one.

Morisot, the person who was also the painter here, was one of several children of a well-to-do family. She did not have to fight her family or leave them to become a painter; they encouraged her. She did not have to deal with "male chauvinist pigs" to gain the approval of her peers; mostly they accepted her. However, she did have to put up with critics who were eager to make what they could of her presumably insufficiently artistic nature as a female. For example, it was suggested that she had learned this or that from her husband's brother, who was also a painter (Edouard Manet). The same critics did not look for her influence on him.

A few years before her death in 1895, Morisot "bemoaned the injustice of her situation.... 'I want to do my duty [to my art] ... I would like others not to make that too difficult for me. I don't think there has ever been a man who treated a woman as an equal, and that's all I would have asked, for I know I'm worth as much as they.'" She also wrote some somewhat different "observations about women's nature and ability: 'The truth is that our value lies in feeling, in intention, in our vision that is subtler than that of men, and we can accomplish a great deal provided that affectation, pedantry, and sentimentalism do not come to spoil everything.'"[6]

Late in her short (by our current standards) life of less than fifty-five years, Morisot wrote of her aims: "Mine is limited to wanting to *capture something that passes*; oh, just something! the least of things."[7] Whatever her differences sourced in "woman's nature," and whatever her limitations and their partial source in patriarchy, she succeeded in capturing many of the "somethings" that pass. We who look at her work now should keep our eyes on the Nows that she gave us and not on the less important things between them. To do otherwise is to risk diminishing her still-living work by sentimentalizing its creator.

# Chapter 10

# Women Have Real Lives of Their Own: Edgar Degas

Edgar Degas did not share the interest of his contemporaries, the Impressionists, in direct painting of outdoor scenes. He "believed that a work of art stemmed from the painter's imagination and should be done from memory in the studio."[1] Degas' vision was aided by his intuitive empathy, which enabled him to portray people as actively engaged in making their own lives. This visceral insight can be seen in his paintings of women. As a result, those works embody some newly changing beliefs about women.

Our look at his work here will be limited to his painting of working women whom men found attractive in one way or another. These were women who worked by dancing, by doing laundry, and (to put it baldly) by "doing" men. Degas' approach differed from that of painters such as Manet and Renoir, who were among the first to paint real young women as alluring sex objects. Degas shows working women as something else. The women in his pictures are as much subjects—in the sense of being active, motivated persons who experience and make their own lives—as are men.

Such a view of Degas' work is unlike what has been expressed by others. The author of a book about Degas' friend and fellow painter, Berthe Morisot, writes that despite Degas' special place in her life, he was a woman-hater.[2] The author of *Idols of Perversity* notes that a critic "summed up Degas' attitude toward women with remarkable acuity. In Degas' work ... woman 'is transformed into an illogical creature ... Degas has pushed to its limits an artistic formula which ... depends upon sentiments of hatred, of aristocratic distaste, and perhaps on a kind of tenderness mixed with revulsion....'"[3] Such views are simply not supported by the artist's pictures that are examined here.

An example of Degas' work with dancers is *The Rehearsal of the Ballet Onstage*, 1873–74 (Figure 10–1). As you can see, "Degas did not picture the ballet only in its floating ecstatic moments. His trenchant eye observed, with devastating realism the moment of preparation in the classroom and the actions of the dancers...."[4] Those preparing to perform, on the left and bottom, include one tying her toe shoes and one who is just yawning and stretching. The diagonal line of six or so are rehearsing what will seem an effortless achievement of artists who model graceful weightlessness. They will also learn to ignore aching backs and limbs, and to disregard bruised feet. At the right in the picture, we see two seated "privileged" men who may be enjoying the rehearsal. More surely, they hope to enjoy the young women.

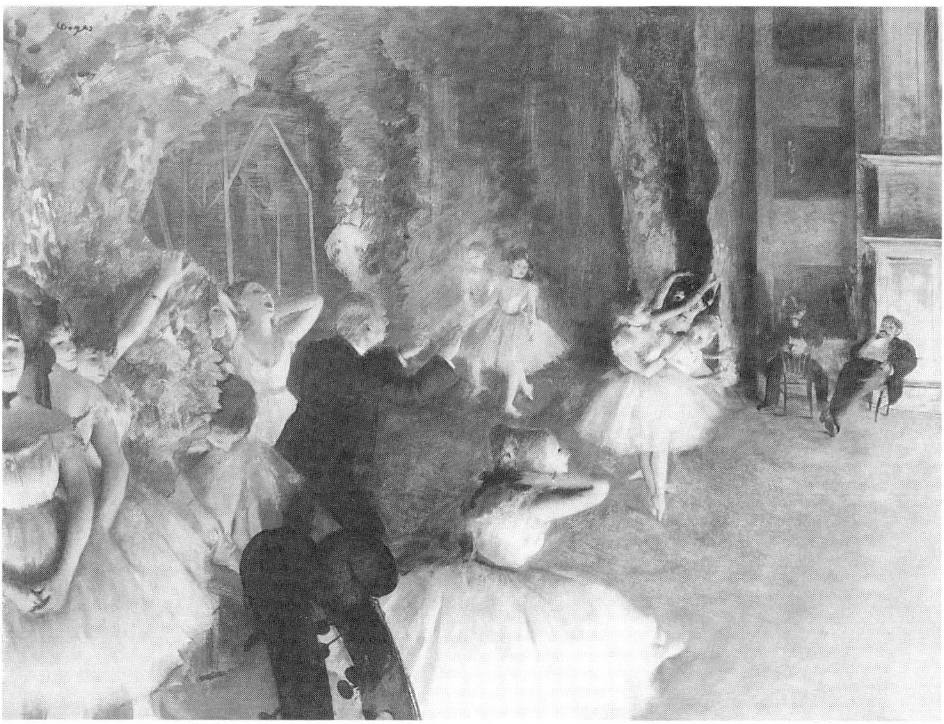

FIGURE 10.1. Edgar Degas (1834–1917). ***The Rehearsal of the Ballet Onstage***, **1873–74. Oil, watercolor and pastel over pen-and-ink on paper, mounted on canvas. 21⅜ × 28¾ in. (54.3 × 73 cm.)** 

Who were these young women? "In addition to sexual allure, exhibitionism, and acknowledged male proprietariness, two other ... aspects of the dancers' lives are frequently noted in contemporary references: their alleged low parentage and their reputed childhood poverty."[5] Much of that assumption — which in the late nineteenth century suggested sexual promotion of girls from a poor background — was the result of moneyed-class uneasiness about the expanding, more visible working class. The young women were in fact workers who had "entered [the ballet world] between the ages of seven and eight.... At ten ... they were given [a reasonable salary] and they spent the next eight or nine years taking classes and preparing to become members of the chorus...."[6] Even before they entered their twenties, the girls realized that their working lives as dancers would not last. They knew that few of them would hold their jobs once they got well into their thirties. Thus, for them the ballet was more than present work and artistry. Sometimes it was "a means of upward social mobility through ... the contacts ... with upper-class men.... Ballerinas were considered available and were patronized by [moneyed men]."[7] We know that some of the sensible young women took economic and social advantage of contact with the affluent older men, just as some of the latter took sexual advantage of some unthinking or ignorant girls.

One response of the middle and upper classes to the expanding, more visible poor and working class was to degrade its young women by sexualizing them. Commercial laundresses could be seen everywhere on the streets. They carried baskets of dirty and clean washing from and to the homes of men with money and families. Such a washerwoman, despite her hard work for perhaps a dozen hours daily, was often regarded as "a brute, if sometimes coquettish, sexual animal."[8] Degas made many pictures showing laundresses. He succeeded in painting them neither as coquettes nor as sexual animals, but rather as working human beings.

He also made many pictures of women whose work (although not apparent to us now in the paintings) was to provide sexual gratification for their clients. These pictures show unclothed "women washing or drying themselves.... The room observed contains some combination of an upholstered chair covered with a towel or dressing gown, a tub, a dressing table, a water pitcher, a bed and occasionally a curtained window."[9] One such picture is *The Tub (bathing woman)*, 1886 (Figure 10–2). This and other nudes he portrayed "are never voluptuous.... They are first and foremost studies of the human form and only secondarily [of] women."[10]

Degas' contemporaries knew what may not be evident to us. A portrait of a naked woman near a tub was not just a picture of a working-class woman, and certainly it was never one of a woman of the leisured

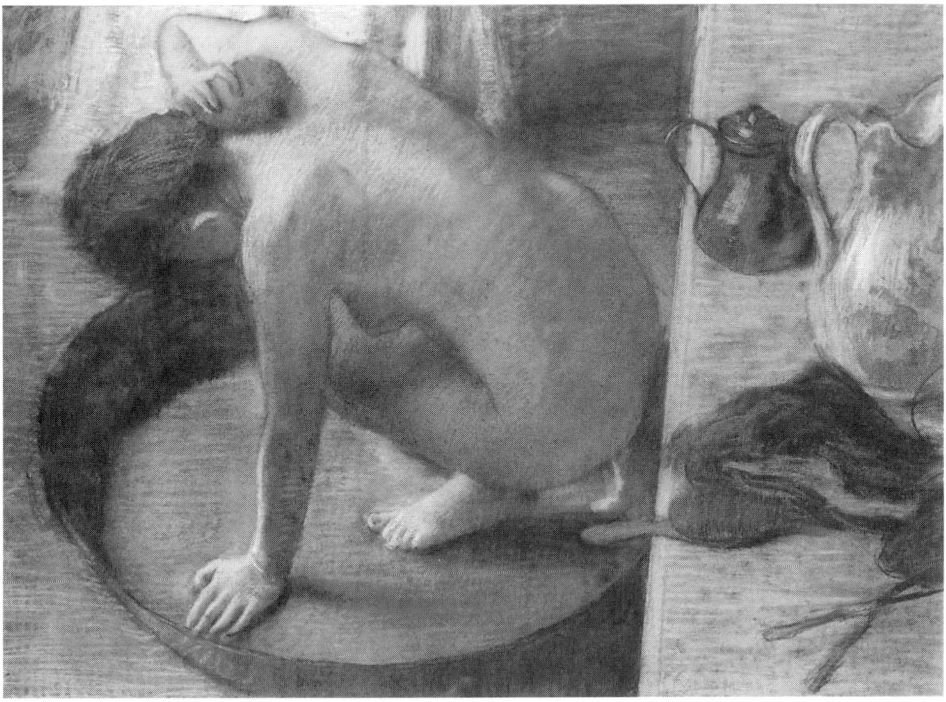

FIGURE 10.2. Edgar Degas (1834–1917). ***The Tub (bathing woman)***, **1886. Pastel. Musée d'Orsay, Paris, France. PHOTOGRAPH COURTESY OF ERICH LESSING/ART RESOURCE, NY.**

classes. Indeed, if he had chosen to portray an unclothed model in a situation suggesting a moneyed-class woman, he would have suffered high heat from critics. The bared bodies of such women were not an acceptable subject — it hinted at something best ignored, that upscale women might have sexual desires. One prominent doctor at that time believed that "The sexual instinct in the civilized woman is ... tending to atrophy."[11]

The bare women Degas portrayed near their tubs were to be found near the bottom of the social scale, since they worked in organized places of prostitution. Each bathed to give her new client the comforting idea (hardly accurate regarding venereal disease) that she was safely clean in general and, in particular, cleansed of her prior customer's penetration.

Despite all that, Degas managed to portray this sex-ridden situation without making it erotic, let alone pornographic. Part of how he did this was by not showing much of the woman's face and, in particular, not painting her to be making eye contact with the viewer. Degas also took care not to show the tub-bather's crotch, thus denying the male viewer the opportunity to make the usual association between the sight of sexual parts and the likelihood that their possessor is sexually receptive. On the other hand, he did not estrange the bare bather by portraying her as a remote unclothed pink female "humanoid," as Cézanne did with *Large Bathers II* (Figure 13–2) and his other bather pictures. Degas brought his empathy to his painterly skills by making each woman cleansing herself neither more nor less than a human being, a person bathing.

Degas also painted pictures that were obviously of prostitutes in a bawdy house, such as *Party at the Brothel*, 1878–79 (Figure 10–3). About this picture, Renoir wrote, "When others paint a bawdy house, the result is usually pornographic ... Degas is the only painter who can [provide] a

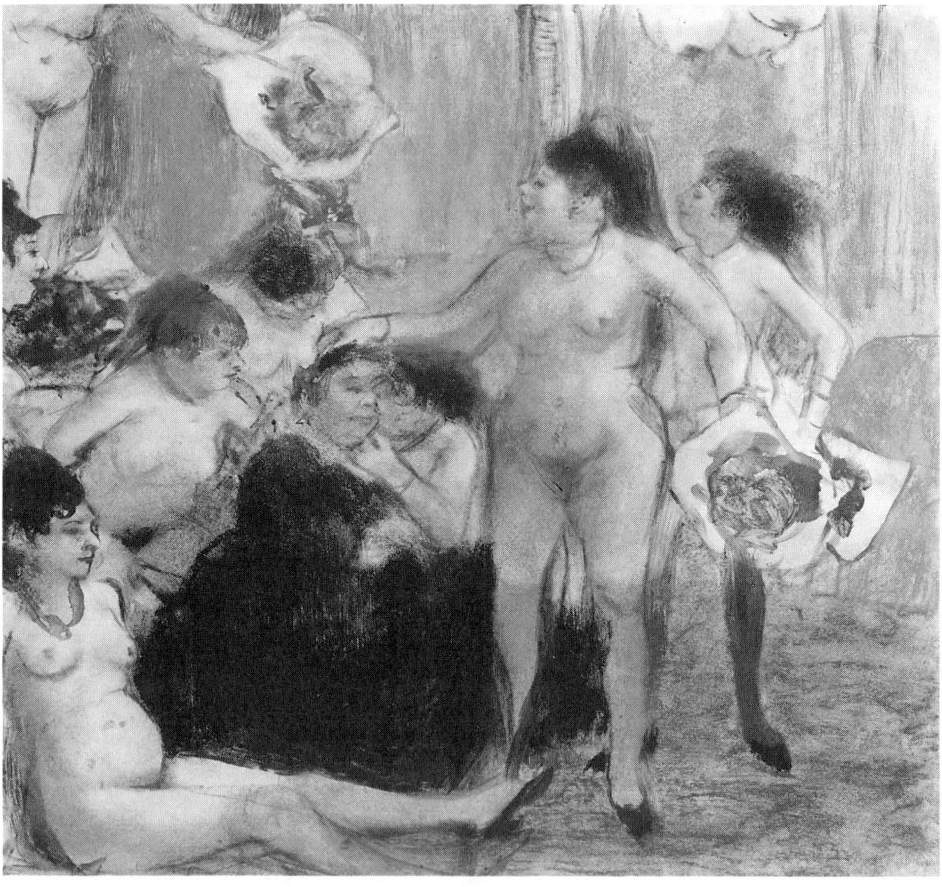

FIGURE 10.3. Edgar Degas (1834–1917). ***Party at the Brothel***, 1878–79. Pastel. Photograph by R.G. Ojeda. Musée Picasso, Paris, France. PHOTOGRAPH COURTESY OF REUNION DES MUSÉES NATIONAUX/ART RESOURCE, NY.

certain joyousness in a subject of that kind."[12] Given his choice of subject matter, Degas does not avoid frontal nudity for one or more women; he confronts it. Thus he portrays female hairy private parts. That is usually enough to make a picture arouse salacious interest in a male viewer, and thus be regarded as erotic or pornographic. How does Degas avoid that? He depicts prostitutes who, if not yet fat and old, are clearly possessors of chunky bodies that are no longer young. He was aware, as we all are, that something has endowed the males of our species with the notion that females who are no longer slim and young are just not as sexually attractive as they once were. (Here is a free-of-evidence "Darwinian" it's-all-in-the-genes story: Younger women are both slimmer and potential bearers of more children. So men with genes that pushed them to select younger women had more children than men without such genes. Over time, the latter disappeared.)

It won't do for someone to protest that it was not Degas who made the women in the picture overly weighty and underly youthful, that was simply the way the models were. That view makes the artist no more than a passive painter of a scene that happened to be available to him. The facts are that Degas created the scenes he portrayed. There were lots of brothels in the Paris of the 1870s. He skipped past the young slim women in them so that he could select older women who had become weightier. While waiting for customers, bawdy house women generally "wore some form of clothing,"[13] so even the nudity of those in *Party at the Brothel* was the doing of Degas. He probably paid them for their time spent posing naked for his sketches. Then, back in his studio, he painstakingly put his vignettes together. In this case, he created a happy little celebration for the boss-lady's birthday.

In pictures like these, Degas shows us that graceful dancers, burdened laundresses, bathing harlots, and past-their-prime prostitutes all have — apart from what they do for moneyed men — lives of their own. The picturing of ordinary lower-class women as erotic objects was itself new in the last third of the nineteenth century. So was Degas' view of working women as fully human beings.

In that they reflect different beliefs, aren't those two kinds of pictures about women, (1) as sex objects and (2) as much fully human as their male oglers, in some struggle with each other? Undoubtedly, they are in conflict. Does the enhancement — by portraying it in pictures — of the belief that women are equally human with men suggest that men should reduce the treatment of women as sex objects? That's hard to deny. However, the stubborn facts are as follows. In the Western world, the twentieth century brought *both* increased enfranchisement of women and massively more marketing of them as entities for the sexual pleasure of men. (For that matter, it also brought gargantuan growth to a related industry, that of beauty products for women.)

All that is the case because, basically, our human heads are occupied by an unruly and conflicted Confederation of Mind rather than a United States thereof. More specifically, it is true that all of us human beings frequently treat others as mere objects— it is part of our nature as subjective beings who actively seek gain for ourselves. The *elimination* of such narrow self- serving at the expense of others may be no more than a wild-eyed dream. However, *limiting* our abuse of others has been a hard-earned part of what we call the growth of civilization.

## *Chapter 11*

# Body and Soul Unreconciled: Vincent van Gogh

Look at what many regard as Vincent van Gogh's first great painting, *The Potato-Eaters*, 1885 (Figure 11–1). "[The pictured people] are peasants, brothers of the miners. Worn out at the end of a day of toil, greedy and uncouth, sitting around the table, they hanker after the steaming dish of humble food."[1] He paints them as crude, almost ugly people. The child in the foreground is the only one of the five whose face might be good to behold. But the artist posed her facing away from us. They are so involved with their food that it is tempting to think that's all they are concerned about. They look as though they live by potatoes alone.

Vincent thought otherwise. Here's what he wrote to his brother, Theo: "You know ... that one of the roots of ... the whole Bible's fundamental truth is: 'From darkness to light....' Experience proves that those working in the dark, in the middle of the earth, as the miners in the black pits, are easily moved by the word of the Gospel and believe it with no effort."[2] No doubt, he was right about them. Like many of the rest of us, they would not have lived by bread or potatoes alone; they also lived by the word of God.

Van Gogh did not live by the material world alone; the spiritual life was important to him. However, he did not believe the world of spirit took the shape of a supernatural realm with God at its center. Thus he described his picture with immense yellow swirls in the sky, *The Starry Night*, 1889 (Figure 11–2), "as an attempt to reach a religious viewpoint without God."[3]

The intimate nearness of the brilliant sky in *The Starry Night* reminds us that, for most of our time on this planet, we thought we lived (on a flat

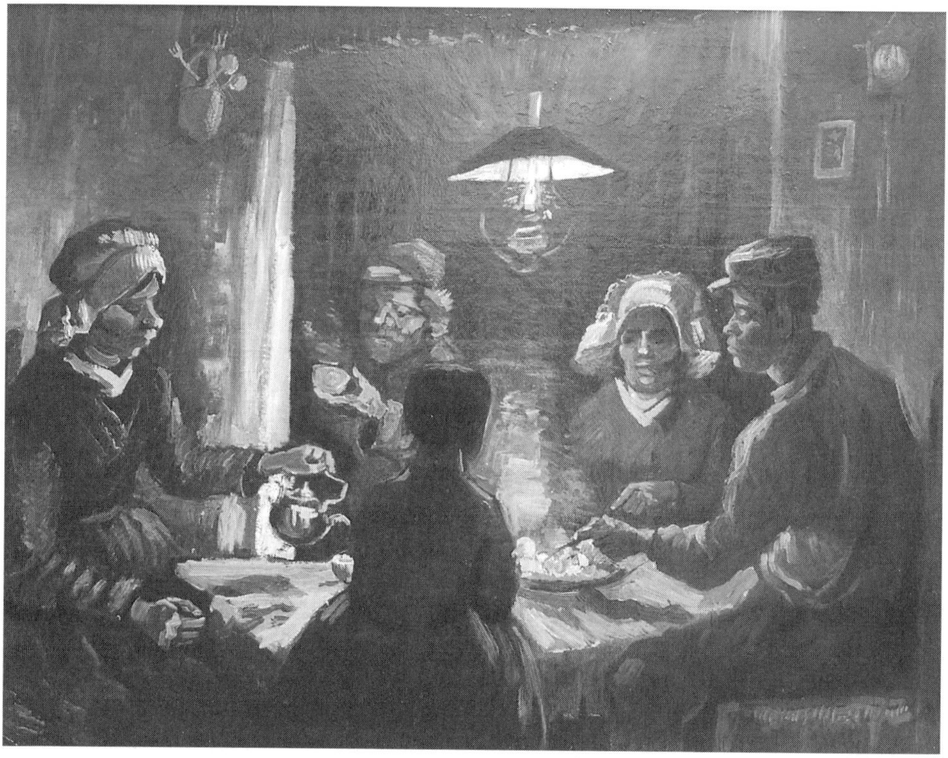

**FIGURE 11.1. Vincent Van Gogh (1853–1890). *The Potato-Eaters*, 1885. Rijkmuseum Kroeller-Mueller, Otterlo, The Netherlands.** PHOTOGRAPH COURTESY OF ERICH LESSING/ART RESOURCE, NY.

earth) below a curved solid dome like "the biblical term, 'firmament.'"[4] The earth was not conceived as a sphere until Greek philosophers developed that idea a few hundred years B.C., and the true size of that sphere was not known until a ship in Magellan's fleet returned in 1522. The earth remained the middle of all things until Copernicus in 1543 replaced it with the sun centering a somewhat greater sphere of the heavens. Until the eighteenth century, the heavens were thought to consist of the solar system with a narrow shell of stars not far behind it. Our idea of the heavens grew rapidly as science and technology continued to shrink our own relative size, but the colossal step in the size of astronomical space did not occur until a few decades after van Gogh's *The Starry Night*, when a large new telescope allowed Edwin Hubble to theorize its immensity. It turns out that twentieth-century knowledge about the universe expanding for billions of years brought a new problem to astronomy. It was called the Big Bang. For

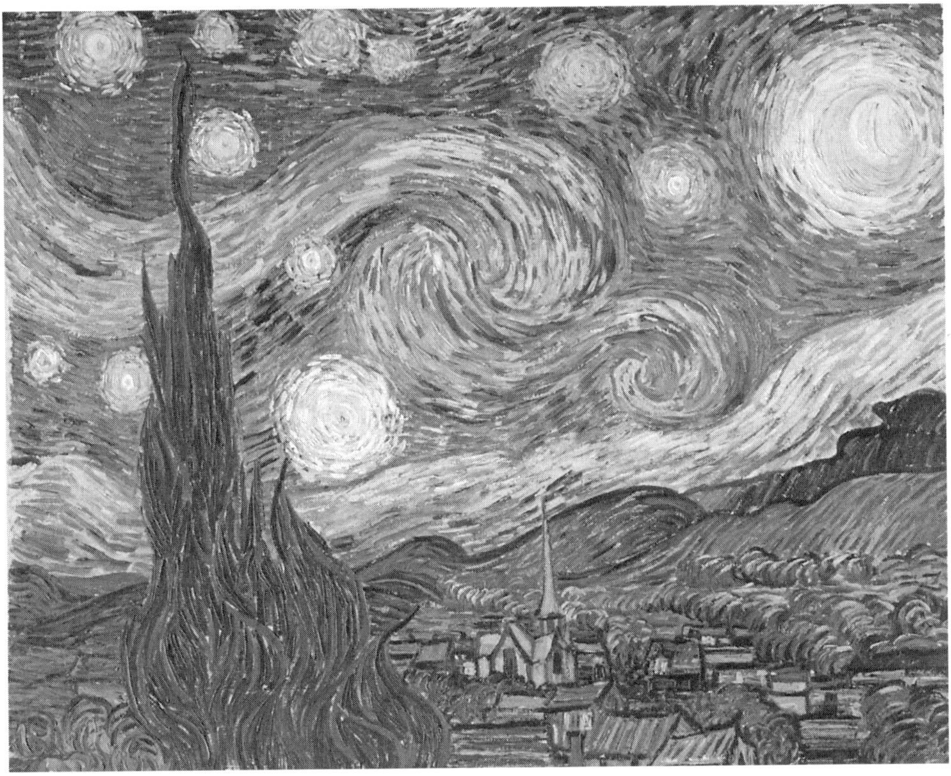

FIGURE 11.2. Vincent Van Gogh (1853–1890). ***The Starry Night***, 1889. Oil on canvas, 29 × 36¼" Acquired through the Lillie P. Bliss Bequest. (472.1941). The Museum of Modern Art, New York, NY, U.S.A. PHOTOGRAPH COURTESY OF DIGITAL IMAGE ©THE MUSEUM OF MODERN ART/LICENSED BY SCALA/ART RESOURCE, NY.

some thinkers, there is "a kind of religion in science; [the belief that] every event can be explained in a rational way as the product of some previous event.... [This faith was] violated by the discovery that the world had a beginning under conditions [where] the known laws of physics are not valid, and as a product of forces or circumstances we cannot discover.... The scientist's pursuit of the past ends in the moment of creation.... This is an exceedingly strange development, unexpected by all but the theologians. They have always accepted the word of the Bible ... God created heaven and earth."[5] Although that biblical origin of everything was said to have occurred only a few thousand years ago, such a recent origin does not have to be viewed as contradicting what scientists see as evidence of far greater age. After all, the scientific evidence could have been "planted" by God for his own inscrutable reasons. In any case, the uniqueness of an abrupt beginning remains problematic for science. Some scientists continue to seek an origin that was part of something continuous.

Turn now to writer René Huyghe's intense prose about this artist's struggle with materiality and spirituality: "Few artists have felt and expressed, as van Gogh did, the pathetic situation of [the] human condition.... All suffer from man's duality; man hanging, torn, between two poles.... This is man's unsolvable paradox. Being of the flesh, he is deeply etched in the physical world and its sorrows; he bears the brand of all that is matter, he is an accident which happened in space and time, ... decay is his only short reprieve from nothingness. But man is also a being of the soul [with] a yearning for the absolute; if he can neither achieve it nor live it, he can at least conceive it and be obsessed by it."[6]

Van Gogh's nature was to be obsessed by the conflict between body and soul. Yet virtually all of us are vitally interested in that struggle [seen here as the conflict between (1) the outer world of material and information, and (2) the inner one of feeling and experience]. This is also a part of the conflict which the art of the last century-and-a-half pictures for us. Vincent van Gogh was one of the soldiers fighting to defend private experience and spiritual belief against the ever-expanding facts and information that were changing the world.

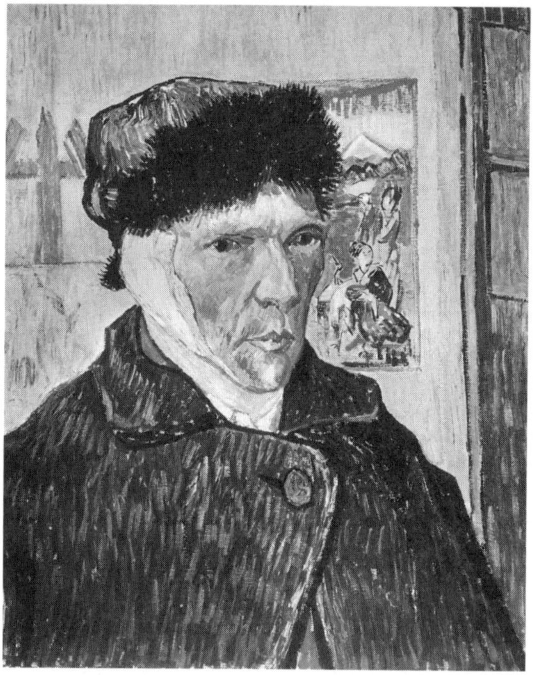

FIGURE 11.3. Vincent van Gogh (1853–1890). ***Self-portrait with cut-off ear and bandage***, **1889. Canvas. Courtauld Institute Galleries, London, Great Britain.** PHOTOGRAPH COURTESY OF ERICH LESSING/ART RESOURCE, NY.

In another letter to his brother, Vincent van Gogh wrote, "Instead of trying to render what I see before me, I use colour in a completely arbitrary way in order to express myself powerfully."[7] In this sentence he expressed his interest in using vivid color to signify more intense feeling. In yet another of his many letters to his brother Theo, Vincent wrote, "I retain from nature a certain sequence and a certain correctness in placing my tones. I study nature so as not to do foolish things—however I don't mind so much whether my colour corresponds exactly, as long as it looks beautiful on the canvas ... to start from one's palette, from one's knowledge of colour harmony is quite different from following nature obsequiously."[8] This trend toward color and subjectivity, later expanded by the artists who were called Fauves ("wild beasts") and Expressionists, was part of art's increasing turn away from the ever-expanding world of material.

"By the end of this 19th century which ... left to his individual loneliness any man concerned about his fate and his reason for living, van Gogh, man of passion and turmoil, could only live his venture by separating himself from the others and sinking into himself."[9] That sentence is a kindly way of saying that van Gogh was something other than most of us. He was a loner — and sometimes (to put it unkindly) he was a loon. To see the loon, or at least the result of a loony impulse of his, look at *Self-portrait with cut-off ear and bandage*, 1889 (Figure 11–3). He painted it not long after he walked to a local brothel and gave someone working there a part of his ear that he had severed.

History has no unbiased account of the events that led to that self-portrait. We know that his sometimes-friend and fellow painter Paul Gauguin arrived a year earlier to work near him. The following is based in part on written material left by Gauguin (who sometimes altered and even reversed the truth about events): "Enthusiasm is soon replaced by quarrels. Van Gogh finds out that the ego remains incompatible, and even incommunicable. [Some time later, van Gogh] throws his glass into Gauguin's face. The next day, as Gauguin is walking in the street, he hears rapid footsteps behind him, turns around and beholds Van Gogh, holding a razor and marching on him. Gauguin stares him to a halt; then, Van Gogh runs away. He runs home and, back in his room, he cuts off his ear with the razor, wraps it up in a handkerchief, and goes out to offer it to a brothel's wench."[10]

A year after van Gogh cut loose part of his ear, "he shoots himself in the chest.... [Days later] complications set in and Vincent dies. He had scribbled these last words for his brother Theo: 'For my own work I am risking my life, and my mind is half gone.... But what do you want?' ... Thus was the life of van Gogh. Absolute failure for him [would become] sublime success for us, but he never knew it...."[11]

In this book, art is *defined* as that which most of those in the art community accept as art. Specifying art in that way has the virtue of being useful. One obvious shortcoming of such a definition is that — as his pictures at that time did not get much interest nor did they get sold — it does not make van Gogh an artist while he was alive. After he was gone, the huge popular turn his art took may or may not have been known to his spirit, but it was widely known here on earth. Now, his art is broadly appreciated by those who display, view, teach, and buy it. Also, van Gogh's "passionate yet controlled deformation of nature and intensification of natural colour in order to create a powerfully communicative art — these were the immediate models for twentieth-century painters seeking expressive means."[12]

## *Chapter 12*

# Spirit and Matter in Tahiti: Paul Gauguin

Paul Gauguin's *The Agony in the Garden*, 1889 (Figure 12–1), can be compared with Delacroix's *Christ in the Garden of Olives* (1827) on the same theme (Figure 5–3) to see some of the ways that painting was changing. The earlier work is realistic in the sense that it looks much like a photograph of posed models would look. The later work could never be mistaken for a photograph, although it did make use of the features of the face of a highly valued model, the artist himself. In fact, Gauguin was a man who "frequently compared his own trials with those of Christ."[1]

A book about Gauguin discusses the changes in painting that make the two same-subject pictures so different. "Symbolist painters must be consistent, [Gauguin] argues, in their violations of anatomy, contour and perspective. Like Van Gogh, who claimed that spontaneous drawing and arbitrary color were superior to 'delusive precision,' Gauguin rejected 'that stupid precision that chains us to material reality. He much preferred mystery and what he called ... 'intense emotions' and 'thought [that] comes up like lava from a volcano.'"[2]

Gauguin's version of *Christ in the Garden of Olives* illustrates the turn away from the objective and toward the subjective in the latter part of the nineteenth century. When he was working on this picture, he wrote, "Art is an abstraction.... Draw art from nature as you dream before it, and think more about the act of creation than the final result; that is the only way to ascend toward God — to create like our Divine Master."[3]

An art critic in 1890 "argued that 'progressive' artists had replaced the Impressionist idea of 'nature seen through a temperament,' with a [theory

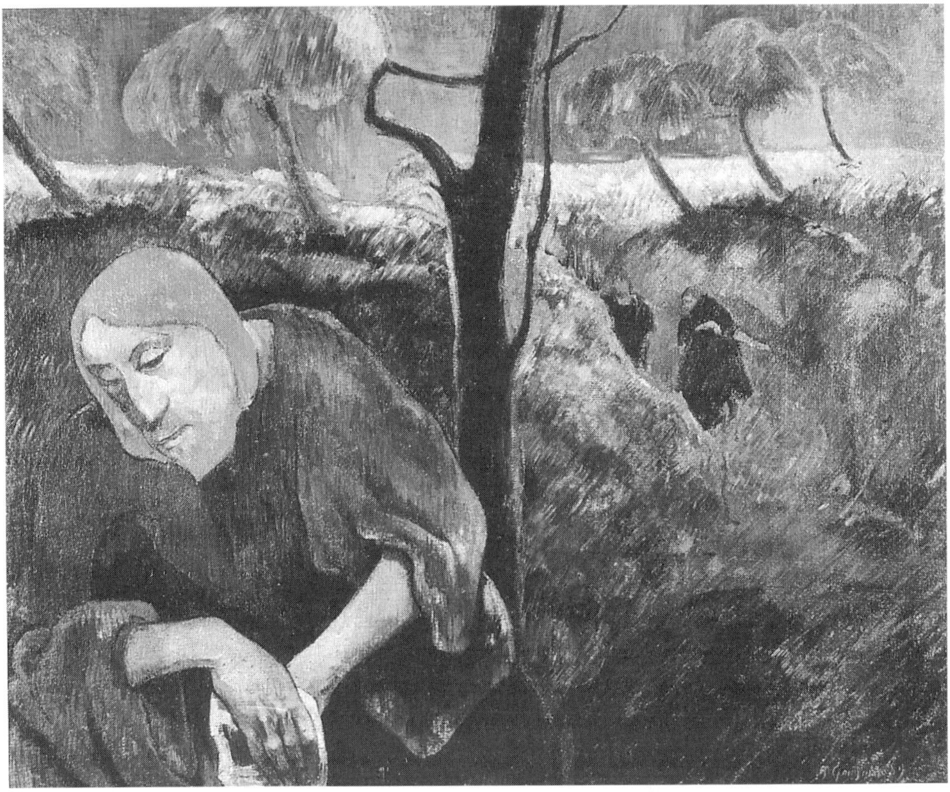

FIGURE 12.1. **Paul Gauguin (1848–1903). *The Agony in the Garden*, 1889. (Oil on canvas). COURTESY OF NORTON GALLERY, PALM BEACH, FLORIDA, USA/BRIDGEMAN ART LIBRARY. PHD334.**

that called for the use of] distortions, signs or symbols to produce pictorial *equivalents* for their *emotions or feelings.* [The new theory] privileged the artist's *feelings* as the source of meaning of the work.... 'Art is no longer just a visual sensation that we set down, a photograph, however refined it may be, of nature.... Art, instead of being a copy, became the subjective deformation of Nature.'"[4]

Certainly, art is the result of both the artist's subjective knowledge (of his ideas) and objective knowledge (of Nature). But both come first through the filter of the facts and spirit of the times. If Gauguin were a God-made clone of someone who lived a hundred or a thousand years earlier, the knowledge of that prior identical twin would surely have been quite unlike that of Gauguin.

Before he went to Tahiti, Gauguin painted in Brittany, where he could "avoid reckoning with the forces of modernity that had made life in Paris with his wife, Mette Gad, so very dispiriting: a depressed market-place that reduced artists to mere hawkers and peddlers of unwanted fancy goods, and an institution of marriage that was 'nothing other than a sale of women....'"[5] The modern world that Gauguin sought to retreat from was one of the results of progress in science and technology. Its newly larger cities filled with an expanding population made it possible to have new marketplaces where new fancy goods, such as Gauguin's paintings, could be peddled.

"[In 1890, Gauguin wrote to a friend,] 'I shall go to Tahiti, a small island in Oceania, where material life can be lived without money.... In Europe, terrible times are in store for the generation to come: a kingdom of gold will prevail. Everything is rotten, man and art alike...."[6] We can quite safely say that Gauguin disliked and disdained the brave new European world that progress had brought. As for things being rotten, it is also safe to say that Gauguin was sometimes an abusive rotter to his wife, as even his son Emil knew. As an adult, Emil remarked that when "I was ten years old ... I saw my father bloody my mother's face with his fist."[7]

The painting *Manao Tupapau* (The Spirit of the Dead Watches), 1892 (Figure 12–2), is one of Gauguin's Tahiti creations. The central figure of a bare young woman resting on a bed was modeled by Teha'amana, a barely teenage girl who lived with him. "Gauguin's representation of [her in this picture is] a hybrid image..., at once French and Polynesian. Gauguin appears to have recognized that spirits played a crucial role in traditional Tahitian culture and religion. Indeed, spirits such as the [gleaming-eyed standing figure who is the watching specter behind the young woman in the picture] form one half of what [can be called] the basic 'conceptual antithesis' of Polynesian culture, the other half of which is *noa*, meaning secular, worldly...."[8]

*Manao Tupapau* "had special salience for Gauguin precisely because it represented his dreamed-of Tahitian reconciliation of matter and spirit...."[9] This hope of Gauguin's is of interest to us here because it shows that our Western view, that there *exists* a fundamental split of matter and spirit, is *shared* by Tahiti's culture. This book's account of modern painting can be called a tale of art's picturing the fears in Western society that matter was *winning* its struggle with spirit.

Gauguin's retreat to Tahiti was the result of *his* fears about the success of matter and materialism in his European homeland. "[In 1891 he went] to the South Seas ... to find an alternative to the modern world and 'that long period of aberration caused by physics, chemistry, mechanics and the study of nature....' For him the Impressionists were the true representatives

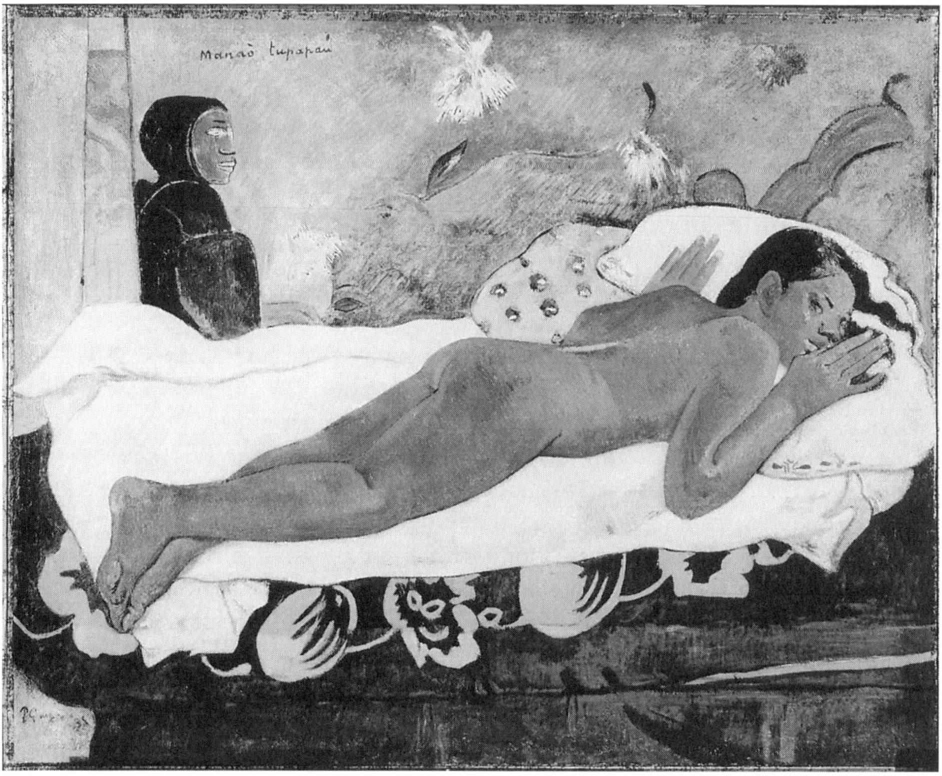


FIGURE 12.2. **Paul Gauguin (1848–1903). *Manao Tupapau* (The Spirit of the Dead Watches), 1892. Oil on canvas. COURTESY OF ALBRIGHT KNOX ART GALLERY, BUFFALO, NEW YORK, USA/BRIDGEMAN ART LIBRARY. PHD764.**

of their decadent century...."[10] Nowadays, no one sees decay or iniquity in Impressionist art, so the notion of such worsening resides in the mind of the beholder. Some of us now find something like the decline and fall of civilization in the art of our own time. About all that, we need say no more than "ho-hum."

Gauguin painted a picture with the stunning title *Where Do We Come From? Where Are We? Where Are We Going?*, 1897 (Figure 12–3). The author of *Gauguin's Skirt* calls it the artist's "modern masterpiece..., a work which recasts Genesis in Tahitian terms."[11] The painter explained that the three questions in the title were related to some of the figures within the picture. The newborn child on the right corresponded to the first question; the central figure plucking fruit raised the second question in terms of the meaning of day-to-day living; the old woman on the left brought to mind the third question.[12] The artist also explained the picture's name as less a

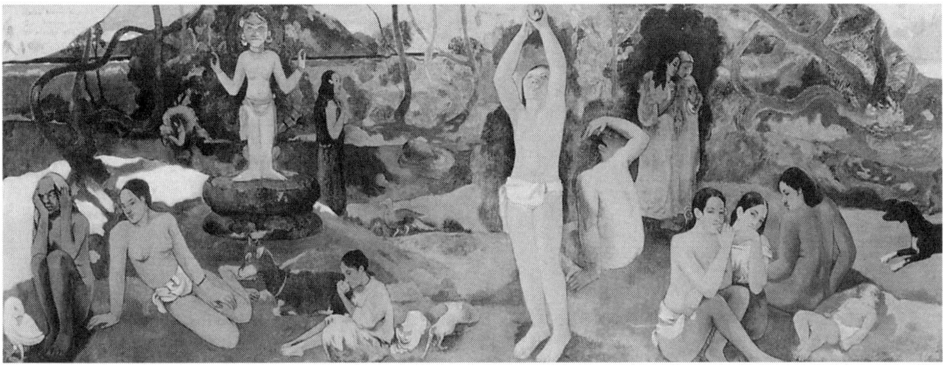

FIGURE 12.3. **Paul Gauguin (1848–1903). *Where Do We Come From? Where Are We? Where Are We Going?* 1897. (Oil on canvas). COURTESY OF MUSEUM OF FINE ARTS, BOSTON, MASSACHUSETTS, USA/TOMPKINS COLLECTION/BRIDGEMAN ART LIBRARY. BST207297.**

title than a "signature," "a philosophical reflection prompted by the picture."[13]

Like some other of Gauguin's works,

> the painting ... features an androgyne, and proposes the correspondence between things visible and invisible, human and divine, and European and Polynesian.... The painting was the result not of cold calculation, Gauguin says, but of an imagination fired from the confrontation with death's abyss. Failures or imperfections in the execution of the work are to be excused—even welcomed, he says—as traces of passion and genius.... Gauguin then ... describes the work, emphasizing its archaistic, fresco-like quality and its magical landscape setting. He also enumerates its ... children, gossips, "figures" [genderless], old women, farm animals, totemic and allegorical creatures, and an idol.[14]

This work is of particular interest here for its title as well as its picture. That title (*Where Do We Come From? Where Are We? Where Are We Going?*) consists of three related inquiries, all of major importance to many people, and all conceptually abstract. Note first that it is exceedingly difficult to convey *abstract* ideas with an untitled painting. (Concepts that are practical are not only easier to make paintings about, but almost *necessary* to make art from paint and canvas.) Handling abstract ideas with paint is as difficult as doing so with wordless gestures. If you were playing the gesture game of charades, your friends and you might spend many maddening and unsuccessful minutes with each of the three questions in Gauguin's above title.

In a letter describing the painting, Gauguin writes, "Behind a tree are two sinister figures, shrouded in garments of somber colour, recording near the tree of science their note of anguish caused by this science itself, in comparison with the simple beings in a virgin nature, which might be the human idea of paradise, allowing everybody the happiness of living."[15] That tells us something we already knew about Gauguin — his belief that science has moved us further from, not closer to, "the happiness of life." Although Gauguin's account does not match very well with his title, that's okay. The title need be no more than a clue to the picture. In a coming chapter we will see that Henri Matisse was wonderfully successful in portraying his ideas of *The Joy of Life* in 1905–6, only a few years after Gauguin's three-question-titled work.

Gauguin's faulting of science—for not delivering happiness to us—is still a commonplace complaint. Like Gauguin, some see such happiness only in the (mythical) *past*, as if it were a birthright of which "we was robbed." Others see it in the future — in the future of this life, in another life reincarnated in new flesh, or in Heaven — where we may finally get what we want and wish we were entitled to. When there's no *pie today*, the idea usually brings up notions of *pie yesterday* or *pie tomorrow*. As we know, the facts of the matter are that science often delivers not happiness, but material progress and better health. When that happens, it turns out that whenever we are not grabbing the goods from the delivery boy, we are content to gripe instead.

"Gauguin copied [for himself] a page from a text [that] describes the mystical union of the gods Ta'aroa and Hina into a single androgynous power: '...this idea of the coexistence of two principles which are God.... One, soul, life or part intelligence of the deity, represented by the name Ta'aroa, is male; the other purely material and consisting in a certain way [of] the body of the same God, is female, called by the name Hina, the two composed, by their union all that exist in the universe.'"[16]

To find the two principles of "soul" and "material," Gauguin could have stayed at home in Europe. Likewise, long before the nineteenth century, European "mythology" would and did suggest that males have more intelligence and soul than do females, whereas females were regarded as closer to the physical world of nature. Indeed, nineteenth-century science played collaborator for such widespread notions by claiming that male brains were superior to those of females.

"During the nineteenth century, most scientists and philosophers were convinced of the incommensurable difference of women and men, and of the inferiority of the former to the latter.... [Some social psychologists] insisted upon the racial and sexual feebleness of non-whites and women...."[17] It may be disappointing to see that nineteenth-century scientists held such views. However, as there's no evidence they were endowed with supernatural abilities, why should we expect them to be unlike the rest of us, who are often content to bow to the common beliefs of the times?

## *Chapter 13*

# An Unearthly Remoteness: Paul Cézanne

"Great revolutionary leaders are men with a single and a simple idea.... [The idea of one such pioneer, Paul Cézanne,] was to see the world, or that part of it he was contemplating, *as an object*, without any intervention either of the tidy mind or the untidy emotions.... Cézanne wished to exclude [the] shimmering and ambiguous surface of things and penetrate to the reality that did not change.... [He was determined] to see the world *objectively*."[1] This artist's "intention was to create an order of art corresponding to the order of nature, independent of his own confused sensations.... Cézanne's heroic attempt ... was in opposition to [what can be seen] in the work of several artists [such as] Gauguin [and] van Gogh [in which we can see a large] decorative element."[2] Those words from art historian and artist Herbert Read provide a good starting place for our look at Cézanne.

To understand the art of Cézanne, here is what may prove a helpful (although over-the-top) way of approaching the timeless, remote *objectivity* of, in particular, many of his later paintings. Consider the possibility that Paul Cézanne was a specimen of *alien UFO-folk*.

With that in mind, reflect on his *Mont Sainte-Victoire seen from the Bibemus Quarry*, 1897 (Figure 13–1). "[One of Cézanne's] favorite themes, ... [his paintings of the mountain have] a mythical grandeur, taking us back to the misty dawn of time.... It seems that toward the end of his life, Cézanne tried to make an ultimate journey back through time, to the birth of the world and original chaos. He said that he wanted his paintings to reveal 'the geological roots of landscape.'"[3] The picture does more than just portray

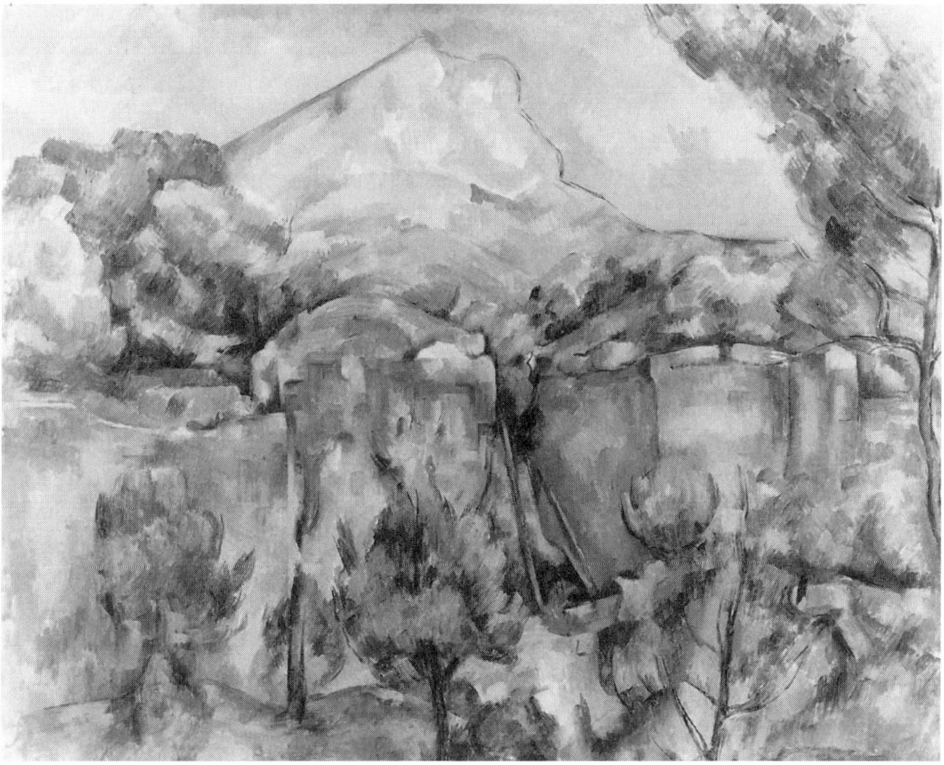

FIGURE 13.1. Paul Cézanne. French, 1839–1906. ***Mont Sainte-Victoire Seen from the Bibémus Quarry,*** ca. 1897. Oil on canvas. 25⅛ × 31½ in. (65.1 × 80 cm.) THE BALTIMORE MUSEUM OF ART: THE CONE COLLECTION, FORMED BY DR. CLARIBEL CONE AND MISS ETTA CONE OF BALTIMORE, MARYLAND. BMA 1950. 196.

enduring forms. It has the appearance of a site never inhabited by any members of our species. Everything in the picture — the mountain, the quarry, the green trees— is remote from the changing world of human sensibility.

Compare it with the famous photograph sent back to earth by the astronauts who went to the moon a few decades ago. Remember the picture, the blue sphere with white circling clouds that is our home planet? That, of course, was not a painting but a photograph, so we would normally think it to be an *objective* document. But photography can also be art, and that picture became art with a significant *subjective* component for all of us. That's because we know it was taken for us by our first intrepid explorers of the moon. You couldn't look at the photo without being aware that it represented the beginning of our efforts to leave our human homeland and explore the untouched universe that surrounds it.

In contrast, Cézanne's painting of Mount St. Victor—for all we know from looking at it—could be one that was made by a non-human visitor from a far planet. Perhaps it is one of a series of semi-holographic close-up photos of our planet taken by the Saucerian occupant of a UFO. Note that there's no reason to be concerned about the seeming lack of reports of inhabited UFOs in Cézanne's time or earlier. In *Alien Dawn*, Colin Wilson writes of "the cosmic expedition [that] reaches Earth (about a hundred thousand years ago) ... to discover ... on the planet ... a type of man that we call Neanderthal.... Sixty thousand years later, [they] pay us another routine visit [and find] our ancestor, Cro-Magnon man...."[4] *UFOs & Alien Contact: Two Centuries of Mystery* tells us that "during the last decade of the nineteenth century ... tens of thousands of citizens reported seeing [an] airship ... typically described as cigar-shaped...."[5] Clearly, whether shaped as cigars or saucers, UFOs have been sighted by earth-folk for a long time.

Turn now to Cézanne's *Large Bathers II*, 1900–1906 (Figure 13–2). As one of his "landscape-figure paintings ... it is composed as a pyramid whose sides are a truncated arch of towering trees and whose base [is a pedestal

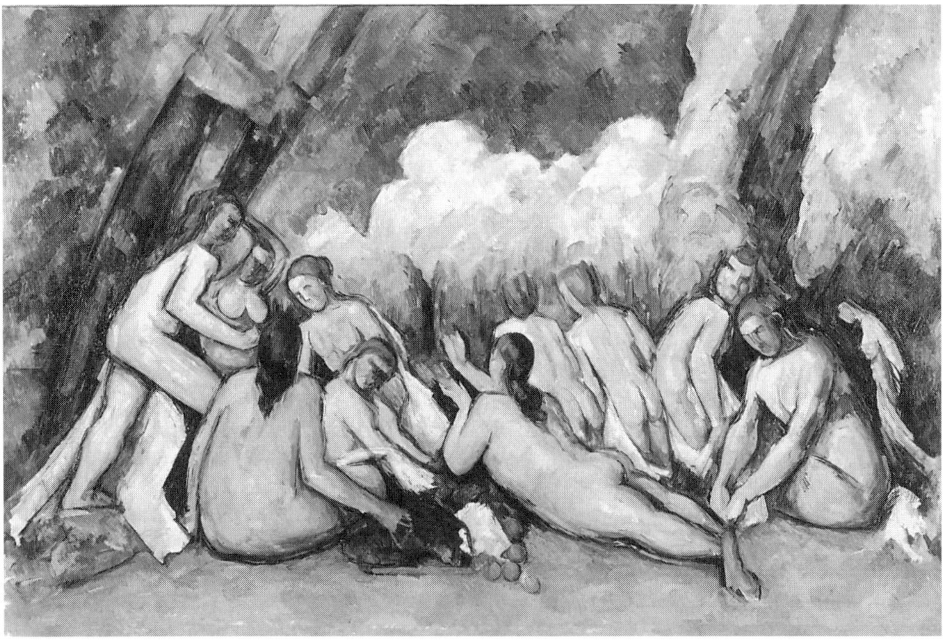

**FIGURE 13.2. Paul Cézanne (1839–1906). *Large Bathers II*, 1900–1906. Oil on canvas, 127.2 × 196.1 cm. National Gallery, London, Great Britain. PHOTOGRAPH COURTESY OF ERICH LESSING/ART RESOURCE, NY.**

of] nudes.... In the foreground, the nude bathers are active, but Cézanne leaves ambiguous exactly what they are doing."[6] Consider the heads of the female figures in *Large Bathers II*. "It was Cézanne's view that the human head, like an apple, could be seen as nothing more than a convenient geometric starting point for a composition in which every element—human or nonhuman—was treated with equal respect and attention. As a result, the most striking feature of Cézanne's ... unconventional portraits is the absence of emotion and personality."[7]

If the ample-bodied unclothed figures show no emotion or personality, and if what they are doing is far from clear, then we should ask—why was Cézanne interested in making the painting? More important, why should we be interested in it? For the first question, a bit of light is provided by considering his (claimed to be human) personal life and its lack of development. "He was past 30 before he found any kind of rapport with a woman outside his family; by early middle age he would no longer paint from a female model, since the presence of nudity made him uneasy."[8] Perhaps the chunky pedestaled women in his *Bather* pictures provided a way of distancing himself while showing that there is more to the world than our own compatibility with it. As for our interest as viewers of *Large Bathers II*, it has the same basis as our interest in *Mont Sainte-Victoire*. There is something transfixing about the picture's remote objectivity. Just why we are gripped by such objectiveness will be examined later.

Cézanne's *Large Bather* women, over a dozen of them with no men in sight, are not, in any significant sense, naked. They are simply without clothing. Here, we may regard Cézanne as an unearthly ethnobehaviorist. Unlike humans, he is a crustacean who has a protectively thick chitinous skin. Periodically he has to discard his crust and replace it with another one that he grows or perhaps finds. *Large Bathers II* is a nonverbal document that he created. It shows that when the strange local crustless creatures are near water, for reasons not yet known, they remove all the artifactual coverings that shield their soft, vulnerable bodies. He will send the record back for further study on his home planet, Sauceria.

About *Large Bathers II*, note also that, in addition to having bare cumbersome bodies in common, the creatures (who happen to be females) all look alike, and all are expressionless. That could be because Cézanne, the Saucerian scientist, has not lived with these peculiar beings long enough to realize that no two are really alike. It could also be that he believes we all (females and males) really are much alike. That's what human ethnologists believe but are not given to saying, as "the psychic unity of humankind is ... anthropology's oldest and most vexing question."[9]

Cézanne shows only that similarities among the females of our species far outweigh whatever differences there may be.

In any case, there is one thing that is clear about *Large Bathers II*. Like Cézanne's many pictures of Mount St. Victor, it shows the remote objectivity that its creator was capable of putting into paintings. "It was the apartness of things ... that drew Cézanne toward his subjects.... Such things endured; they rebuked and put in place the conflicts of his passionate, inhibited character."[10]

To see something of those conflicts underneath his own crust, let's change from the distant view of *Mont Sainte-Victoire* and *Large Bathers II* to a revealing (of himself) look at his *The Robbers and the Donkey*, 1869–70 (Figure 13–3). "[Cézanne] knew Greek and Latin literature.... In all likelihood, he had read the [writers who] tell the story of a man who was changed into an ass and captured by thieves. He demonstrated this knowledge when he painted *The Robbers and the Ass [Donkey]*. But this leaden sky and murky

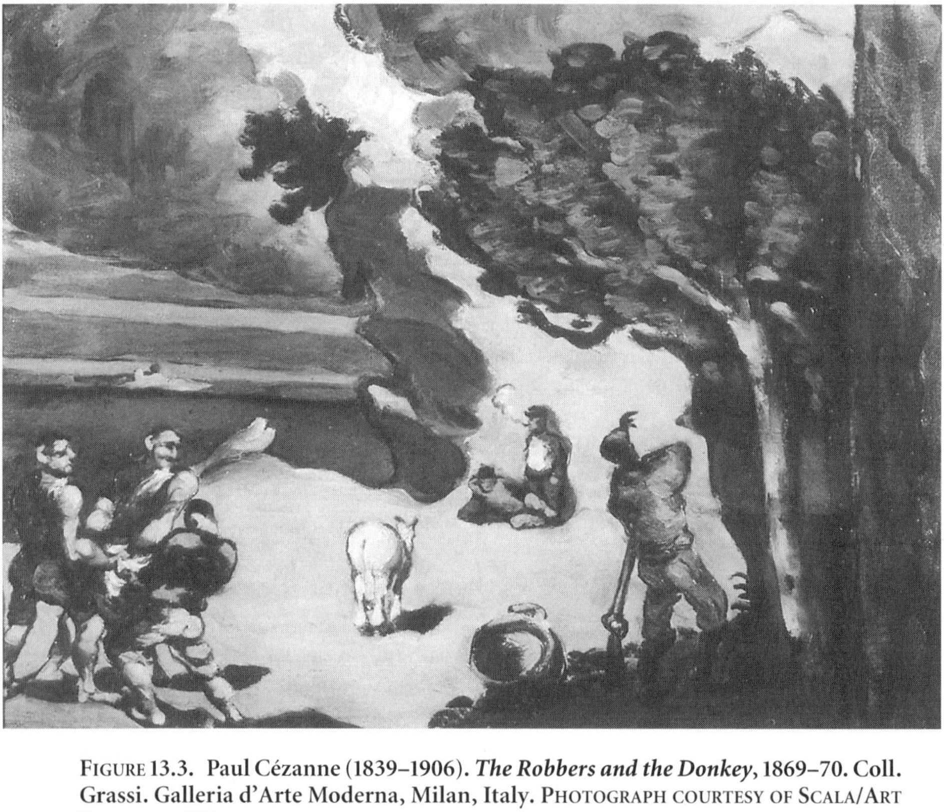

**FIGURE 13.3. Paul Cézanne (1839–1906). *The Robbers and the Donkey*, 1869–70. Coll. Grassi. Galleria d'Arte Moderna, Milan, Italy.** PHOTOGRAPH COURTESY OF SCALA/ART RESOURCE, NY.

sea, this wind rushing through dark foliage, these muscular robbers dressed in red or blue, this strange figure smoking..., this pathetic little white ass ... this is an imaginary world which Cézanne could only have seen in dreams."[11]

Cézanne has seemingly given us an old-fashioned story-picture. Clear enough is the white donkey in the center, and clearly the thieves at the bottom left are about to act. On the far side of the donkey, however, sits a calm observer who seems indifferent to the local action. He looks far away, looks at something outside of the local picture. As it happens, strange Saucerians like Cézanne see themselves with split personalities and separated personas. Each has two entities: one of them is a thick-crusted body. The other, a tender mind-and-spirit, shows itself only when they occasionally shed the thick skin. The ethno-behaviorist from elsewhere has created a picture that shows the local human clodhoppers are quite interested in capturing his encrusted body, shown as a white donkey. As for his thin-skinned mind-and-spirit within the calm observer, the homo sapiens don't care about it, or perhaps they can't even see it.

Whether or not he had genes from out of this world, Cézanne pictured two aspects of himself in *The Robbers and the Donkey*. He was a man more fractured than split by different facets of his personality. Writer John Rewald tells us Cézanne was "withdrawn and shy, was at the same time romantic and boisterous, exuberant and unstable, overconfident and moody, hesitant and stubborn, proud and melancholy, sensitive and insecure, suspicious and irascible."[12] "All his life, he would alternately reach out to people and thrust them away. He did not want anybody, he said, 'to get his hooks into me.' He meant this so literally that he developed a morbid fear of being touched."[13]

The picture suggests that Cézanne knew how different he was from his fellows. He knew that sometimes he was sensitive observer and sometimes unthinking donkey. He found a way to dramatize this knowledge with the old story of *The Robbers and the Donkey*. Indirect support for this view is available from Cézanne's words: "A painter like me, there is only one every other century."[14] There is also such support from the strange reaction to him from a group of children in the following anecdote: "A man in a bowler hat sauntered along the street.... Suddenly he raised his head and looked over his shoulder. It was then that the children started throwing stones at him. One is staggered when one recalls who this man was.... His name was Paul Cézanne."[15]

The question remains: Why then Cézanne's fame and acclaim? In part, it was for the mastery suggested by his later pictures. But it is also true that his work does not show what his fellow painters who were Impressionists portrayed: What it was like for an ordinary person to live in those times. So the other part of his claim to fame is his indifference to the subjective life of engagement with other humans and with the human world. Cézanne painted the world and its inhabitants as uncaring, spiritless objects.

Here is a likely reason why the art community did and does find that remote view of life — which Cézanne portrayed with such skill — to be fascinating and important. In his times, science in particular and objective knowledge in general were beginning to make rapid changes in the way in which people lived their lives. For most of the hundreds of millennia of human existence, we had lived with no more than a few human-made objects— artifacts such as pieces of stone shaped to serve as hand axes. By the nineteenth century, the opposite began to be the norm. A city person could go through a whole day and encounter no more than a few *natural* objects. By then (and far more so now), most everything we had frequent contact with (at least in cities) was human-made, an artifact. Even in Cézanne's still-life pictures, the objects embody "nature as colonized by [humans, the] arrangement thematizes human dominance by reference to manipulation."[16]

So what's the connection with Cézanne's remote objective world? The link is this— we humans expanded our objective knowledge for a simple and powerful reason, to benefit ourselves. We wished to mine the world for more than its precious natural objects. We also sought and found the means and materials to make the countless artifacts that we find useful. It is as if scientists had dug from Mount St. Victor every single thing — store-bought packaged food, manufactured clothes to cover us, sheltering rectangular houses, transporting wheeled machines, you name it — everything that, for most of our millennia, we simply did not have.

During those millennia, we made sacrifices to the countless deities that we thought existed everywhere — we gave in the hope of getting in return more of what was good and less of what was bad. However, the world — represented by the mountain that Cézanne made so many pictures of— turned out not to be some kind of malevolent agent who demanded tribute. Neither was it a beneficent deity that bestowed on us wonderful cargoes of goods in exchange for good behavior. If anything is obvious, it is that Mount St. Victor, like the rest of the objective world, simply does not care about us. In truth, we often take more discomfort than comfort in that, but we sometimes accept it as an inscrutable objective fact. And Cézanne showed us what the fact of an indifferent world looked like.

Imagine there was a prophet — two or twenty thousand years ago— one who foretold countless slowly acquired step-by-step building blocks of science-like objective knowledge. Then that seer foresaw that the resultant vast structure would totally change and in most ways improve the way we lived. Would such a prophet have been honored in any country? An oracle like that would have been rewarded — at best — with discrediting yawns. Yet such change and improvement is what indeed happened. And Cézanne's pictures do indirect honor to the power of the knowledge about nature that we humans mined from the world.

# PART IV

# Science Repels and Attracts: The Early Twentieth Century

### *Chronology of Technological and Scientific Developments*

1900: Planck formulates quantum theory
1903: Wright brothers fly a powered airplane
1905: Einstein formulates Special Theory of Relativity
1907: Pavlov studies conditioned reflexes
1913: Geiger counter introduced
1915: Einstein postulates General Theory of Relativity
1917: Mount Wilson 100-inch telescope installed
1920: Thomson patents submachine gun
1922: Insulin first administered to diabetic patients
1926: Goddard fires first liquid fuel rocket
1927: Lindbergh flies from New York to Paris
1929: Hubble measures red shifts in spectra of nebulae

# *Chapter 14*

# The New Subject Meets the New Object

The most significant advance of science in the nineteenth century was surely Darwinian evolution. In "the long perspective of four thousand years, the Western world has discovered or invented only two master ... narratives of high explanatory power. They affect every aspect of human life today.... The sequence of Hebrew covenants followed by the New Testament Redemption story offers an account of things bestowed on us [by] God. That religious ethos ... had no full-fledged competitor until the elaboration of Darwinian evolution in the middle of the nineteenth century."[1]

Large changes induce large reactions to the source of innovation. The response to the "scientific bulldozer came most markedly in the 1870s and the 1880s, which saw a revival of fundamentalist Christianity.... It was not only Darwinism that was perceived to threaten the mildly religious Victorian world. The upsurge of geology ... fostered antibiblical notions, such as the immense age and gradual development of the Earth. Investigations of life processes, inexorably cutting away the shroud of mystery that had singled out Life as something outside Newton's mechanistic universe, were deeply troubling—and still are, to some people."[2]

The revivers of fundamentalist Christianity perceived the new objective knowledge from science as a powerful threat to religion. They saw how the God of the Bible was being reduced to an almost impersonal Deist God who did no more than start the world. They needed a counter force that would not only be strong, but would also be made of stuff that could stand strongly against the force of the new knowledge.

To get that stuff, they made good use of the Greek metaphysical tradition, which tells us of a split in the nature of the world. "There are two kinds of substance: 'material' and 'spiritual.' Material substances are temporal and divisible.... Spiritual substances are timeless and indivisible.... How do human beings know spiritual substances?... Human beings share timeless immortality with [God] through their immortal soul."[3] Spiritual substances are thought of as part of our "hearts" or our souls. Whatever they are, most of us believe them to be quite unlike material substances. Thus much of the new knowledge about the world from science and technology is thought not to affect them.

"The main targets of the fundamentalist crusade were biblical 'higher criticism,' which distinguished between spiritual [validity] and scientific truths in the Bible...; Charles Darwin's theory of evolution...; and the theology of a 'social gospel,' which envisaged the kingdom of God not as a future millennium, but as a socioeconomic struggle.... [Mainline religion was falling] prey to the satanic forces of modernism...."[4]

Most of us would agree that there was and is conflict between fundamentalist Christianity and Darwinian evolution. Yet many of us think religion and science should not be in conflict precisely because the two occupy non-overlapping domains of authority (as Stephen Jay Gould suggests in *Rocks of Ages: Science and Religion in the Fullness of Life*[5]). Why then are there conflicts between them when there "should not" be any? The two clash because, like other kinds of human thought and endeavor, neither is as pure as is sometimes claimed. Their domains are not as separate as their impassioned advocates claim.

What did the late-nineteenth-century revival of fundamentalism (as a response to science) have to do with the subsequent changes in art? The realm of ideas—which we use to think about, describe, and paint versions of our human world—was becoming increasingly split as science expanded. It is difficult to argue with science on its own grounds. Even among the intensely religious, few claim that scientific truths are dishonest. It's easier to quarrel on more subjective, high-principled grounds. And easiest of all is to turn away from the outside world and turn to the inside one of human subjectivity.

In *A Concise History of Modern Painting*, it is put this way. "Two general directions of development gradually became apparent, and no doubt these correspond to the main divisions of human temperament, which we call introvert and extrovert."[6] The subjective (introvert) side of the increased polarization can be seen in the work of new painters called the "Fauves" (wild beasts). "The world presented in the Fauvist landscape [takes two shapes.] It is an immediate world, a picture of life that exactly matches the immediacy of its depiction. Beside this, however, runs a parallel current, which seems at first sight to repudiate the other, [for it dreams] of the happiness of a far-off Golden Age."[7]

Both of these Fauvist trends—showing a vivid present and a dreamed-of golden past—shared the more emphatic new subjectivity. It took another new form of painting, that of the Cubists, to exhibit the new, more polarized objectivity. The Cubists were covert "groupies" in the endlessly changing and increasingly dominant world of objective knowledge. A slogan for the Cubists might have been, "You wanna see real innovation? Just look at us!"

In the following chapters, we'll look first at Henri Matisse's work, which served to widen the subjective side of the street. Then we go across the boulevard to the Cubist product of Pablo Picasso, as it is his Cubist works that expand this story of the increasingly two-sided nature of modern painting. Lastly, in a chapter ahead, we wonder about one weird work of Marcel Duchamp's, one that he did not create but merely labeled. It made viewers ask a new question: What is art—and what ain't?

## *Chapter 15*

# Decorating the Inner World: Henri Matisse

Nowhere in the art of Henri Matisse "does one feel a trace of the alienation and conflict which modernism, the mirror of our century, has so often reflected."[1] Matisse indeed "lived in trying times. Countless strikes, uprisings, assassinations, and two world wars exploded around him. Yet his paintings ... ignored all [that] controversy."[2] That raises an obvious question: How did Matisse avoid all that anxiety about the changing outside domain? Put plainly, the answer is simply this: He turned away from the world and toward imagined and real aspects of life's inner side. Matisse and other artists used the word "expression" to convey some of their feelings about painting. "Art must be expressive of 'a nearly religious feeling towards life....' 'What I am after, above all, is expression.... I am unable to distinguish between the feeling I have for life and my way of expressing it.... The whole arrangement of my picture is expressive.... Composition is the art of arranging in a decorative manner the various elements at the painter's disposal for the expression of his feelings.'"[3] He also said, "Color was not given to us in order that we should understand nature, [but] so that we can express our own emotions."[4]

In *The Joy of Life*, 1905–6, Matisse demonstrates what he meant. That work shows a lovely imagined world painted in attractive unreal colors. It portrays an idyllic scene of women and some men living free and easy. "The theme [is] joyful acceptance of life in a harmonious human setting. This atmosphere of serene joy ... is reflected in an Arcadian scene: groups of figures in paradisiac nakedness, resting, playing, embracing in a sun-drenched landscape.... The artist does not hesitate to deform the natural appearance in order to create the simplest, most expressive figures...."[5]

Living there is not only easy, it is without any human-made objects except those with which they make music. They are living without science, without technology, living even without those ancient artifacts called clothing. *The Joy of Life* shows us times as they are not now, as they never have been, and as they may never come to be.

One of the great breakthrough paintings of the twentieth century—that's what critic Leo Steinberg calls *The Joy of Life*. Then he tells us not only that established artists reacted negatively to it, but also that such nay-saying was the common reaction of the establishment to the new in art.

> It made people very angry. Angriest of all was [painter] Paul Signac. "Matisse seems to have gone to the dogs. Upon a canvas of two and a half meters, he has surrounded some strange characters with a line as thick as your thumb." [Steinberg continues, saying he cites] this affair merely to suggest that Signac [was acting] like a member of [Matisse's] public.... One year later, Matisse went to Picasso's studio to look at *Les Demoiselles d'Avignon* [The Young Ladies of Avignon, Figure 16–2]. The picture, [Matisse] said, was an outrage, an attempt to ridicule the whole modern movement. At that moment, [Matisse] was acting, typically, like a member of Picasso's public.... Such incidents illustrate a general rule that whenever there appears an art that is truly new and original, the men who denounce it first and loudest are artists. The men who kept Courbet and Manet and the Impressionists and the Postimpressionists out of the salons were all painters.... [By] 1912, the triumphant Cubists were to reject Duchamp's *Nude Descending a Stair*.[6]

The nude women in *The Joy of Life* also elicit some comments from Steinberg. "Matisse disturbed certain habitual assumptions. For instance, one had always assumed that [a male viewer] felt entitled to some pleasurable reward if he focused on painted figures, especially if these figures were joyous, female, and nude. But in this picture, if one looks at the figures distinctly, there is a curious lack of reward. There is something withheld, for the figures lack coherence or structural articulation.... Perhaps Matisse was trying to make his individual figures disappear for us...."[7]

Let's borrow an idea about why the individuals in the picture disappear into the group. (This concept also helps to explain the art-establishment practice of regularly resenting the "new guy.") Philosopher Susanne Langer tells us that within our minds "the concept of individuation ... has a converse ... namely, 'involvement.'"[8] In our many-sided human nature, the other side of being a distinct individual is involvement with a group of one's own kind of people. Yes, we are self-centered creatures. However, we also feel ourselves part of a like-minded community.

One thing Matisse's picture suggests is that the happiness of life is best gained *socially*. Emblematic of such happiness is the circle dance of six all-alike young women in the center background of the picture. Likewise, this *communal* involvement is emphasized when we see that all the figures in the foreground are shown to be so much part of the group that they are only marginally differentiated from one another. In fact, anywhere in human life, the group discourages individuation. More accurately, the social mindset of the individuals in a group discourages those individuals from being independent of others. (Putting the emphasis on the individuals avoids muddy-minded making of the group into a "super-organism.")

We can see the same conflict between individuation and involvement in some of the classical ballets. In *Giselle*, the life of the central young woman is ruined when she chooses an outsider, someone who is not part of her peasant group and class. In *Swan Lake*, a similar conflict is enacted when a poor swan and a rich human fancy each other. Related to this community/individual conflict is other evidence that we see in many ballets. They structure activity as alternating between a group—an ensemble of dressed-alike, perform-alike dancers—and a few individuals (or couples) dancing by themselves.

So happiness is to be hoped for in the company of like-minded others of one's own kind. Even cows know that, as anyone can tell you who is familiar with the sport of using a cutting horse to keep a cow away from the rest of the herd for a few minutes.

Despite all the nineteenth- and twentieth-century talk about—and real gain in—individual freedom and its value, we humans are, unlike our cats but like our dogs, a social species. We might never have developed our big brains with wide-ranging minds were we not evolved from social animals. For social creatures, happiness is best found in the pack, whereas unhappiness is not being allowed into the group, or being cast out. Of course, we humans have advanced far beyond cattle or dogs. For human insiders, artists or otherwise, unhappiness is often being confronted by someone who tries to push into the pack with newfangled ideas that disturb the establishment.

Exhibited in 1906, *The Joy of Life* also "became the focus of debate about the value of 'decoration....' For some critics the echoes ... evoked by the frolicking and reclining figures must have jarred with the formal handling of the work. The distorted scale and perspective, the stylized rhythms, loosely painted and flattened areas of colour [were blamed on artist André] Derain and his dreams of 'pure decoration.'"[9] In its meaning of "ornament," *decoration* is not something that is objectively essential. Its appeal is mainly subjective. Decoration is part of Matisse's move toward the inner world. Likewise, he deliberately draws his figures to be unlike the reality that would be seen in a photograph.

The individuals in *The Joy of Life* are almost all young females. That is another part of why the picture works so well. In a man's world (which is a description of perhaps all human social groups in the past, and most in the present), starting with unclothed young women provides a good basis for showing the happiness of life. So the painting is a touch erotic. In *The Nude*, critic Kenneth Clark says, "no nude, however abstract, should fail to arouse in the spectator some vestige of erotic feeling, even though it be only the faintest shadow—and if it does not do so, it is bad art...."[10]

Matisse keeps the eroticism well below the threshold of taking over *The Joy of Life* by providing emotional distance with figures mainly in outline, figures with no depth or detail. However, in his *Blue Nude* ("*Souvenir de Biskra*"), 1907, there is a figure of a young woman (painted in obviously unreal blue) who is lying on her side and facing the viewer. She is more erotic than those in *The Joy of Life*. What makes the Blue Nude erotic? She is not only naked, young, and three-dimensional in appearance but she has a narrow waist with broad hips and sizeable breasts. Evolutionary psychologists (using some science) might suggest the appeal of the latter features has its origin in our genes. A slim waist (which intimates youthful ability to have many offspring) and wide hips with bountiful breasts (which evoke the idea of good infant-bearing and rearing capability) together catch the eye of the Darwinian male, whose genes push him to look for someone who could deliver a big litter.

Darwin apart, why is it that a bare, shapely female body has this effect? It is not simply biological, not just instinctive. It also depends on something *learned* within a society. We know that because in places where women wear little or no clothes (as can be seen with some of Gauguin's South Seas paintings of near-bare young women), *their* men do not find that to be erotic. That's just the way that women dress—or rather, don't dress—in tropical climates. However, wherever the weather is colder, women normally wear clothing. Then, the unclothed body of a real young woman becomes not only bare but *naked*.

It is suggested here that (other than in the Tropics) the absence of clothing is taken by men — below consciousness, no doubt — to be not just erotic in a general sense, but specifically a symbol of the woman's *sexual receptivity*. It is such seeming evidence of willingness to engage in sexual activity that most men, like the males of other species of mammal, are ever on the alert for. Non-human male mammals, however, can count on periodic heat in females to provide strong aromatic or visible evidence. For the human species (which is from a line that abandoned female heat), males are frequently all atwitter in response to unreliable symbols of sexual receptivity, instead of surefire signals that set nostrils or eyeballs aflame.

So men are likely to get excited when they see a woman who seems to be *revealing* herself as naked, and *thus* as sexually receptive. Of course, a painted picture of a nude woman is hardly as stimulating as the real thing. As we know, the frequent opportunity to see many such real things (unless you went to a place where receptive females were for actual hire) did not exist in Matisse's times. It has since become part of the social world in which we (in Western society) live. Although some do, many sexy-looking young women now no longer sell themselves. They sell their live, film, or TV performances as more or less bare erotic objects. They also sell, with bait-and-switch tactics, almost everything else.

In *Primitivism, Cubism, Abstraction*, we learn something about sex and the *Blue Nude* that is not apparent to untrained viewers: "Matisse reworks a set of well-established nineteenth-century conventions; the female nude or 'odalisque' in an oriental setting — often suggesting the harem or prostitution...."[11] The ordinary male viewer does not need this additional knowledge suggesting harlotry to feel a link between nudity and sexual receptivity. Thanks to Matisse's skill, the shapely Blue Nude herself provides the connection.

The author of *History of the World's Art* tells us about Matisse and the Fauves: "About the turn of the century Expressionism made its appearance.... [The new idea] meant that art was no longer content to concern itself with the surface of things; that the impression received ... no longer seemed as essential as the psychic experience to which the artist gives external expression.... [However, in response to a 1905] exhibition at the Grand Palais ... the exasperated critics spoke of cage de fauves, 'a cage of wild beasts,' and the name of Les Fauves had once for all attached itself to the early French Expressionists.... Henri Matisse is the most consistent representative of the movement.... Expressionism became ... an utterance in which the psychic content was more important than any mere formal qualities or laws of composition."[12]

Matisse himself writes that "the Fauves cast the perceptual world into the background in favor of an intuited world. 'This intuition is a consciousness of life which is of the nature of a religious attitude.... The work of art is a sample of the feeling it expresses. It is the only rendition of that feeling in concrete form.'"[13]

Matisse lived long enough to have known the work of Mark Rothko, Willem de Kooning, Lee Krasner, and Jackson Pollock (some of which we'll soon see). He might have regarded them as pushy outsiders trying to get into the art community. He might also have thought that—even as they turned further from the ever-larger fears about the outer world—they brought decorative painting that expressed feelings to a new high point.

# Chapter 16

# Cubism and the Illusive Body: Pablo Picasso

When it came to painting, Pablo Picasso was like Proteus, the Greek sea god who readily changed his appearance. Although he was at the center of the art movement called Cubism, it never contained shape-shifter Picasso. He "either moved in and out of [movements] or simply transcended them...."[1] It is mainly the striking Cubist paintings of Picasso (and others) that relate to this account of art's cracking mirror, so let's try to limit our attention to them. It is pertinent to note that, for all the innovation in Cubist works, human warmth is not to be found in them.

Just look at and into Picasso's Cubist-period *Daniel-Henry Kahnweiler*, 1910 (Figure 16–1). In this portrait of a human being, there's no sign of character or compassion, and not much in the way of a symbol of anything else within our experiencing insides. Of the presence of Kahnweiler (whose gallery sold many of Picasso's paintings), we are given only the slightest evidence, mainly his eyes. The elusive Kahnweiler's presumed body and the rest of the picture are made up of straight lines, a few smooth curves, and muted shaded colors.

What on earth — if the warp of modern times has left us still on earth — has happened to pictures of the changing outside world? No longer do we see new features take a form such as the open brushstrokes by Manet and then by the Impressionists. Now it is the Cubist dismembered planes in vague space that reflect something new. What can it be reflecting? It casts light on nothing less than novelty itself, just like the freshness of ever-changing and increasing artifacts from science and technology. Note that there's no sign that Kahnweiler's is a better world, nor is it shown as a worse

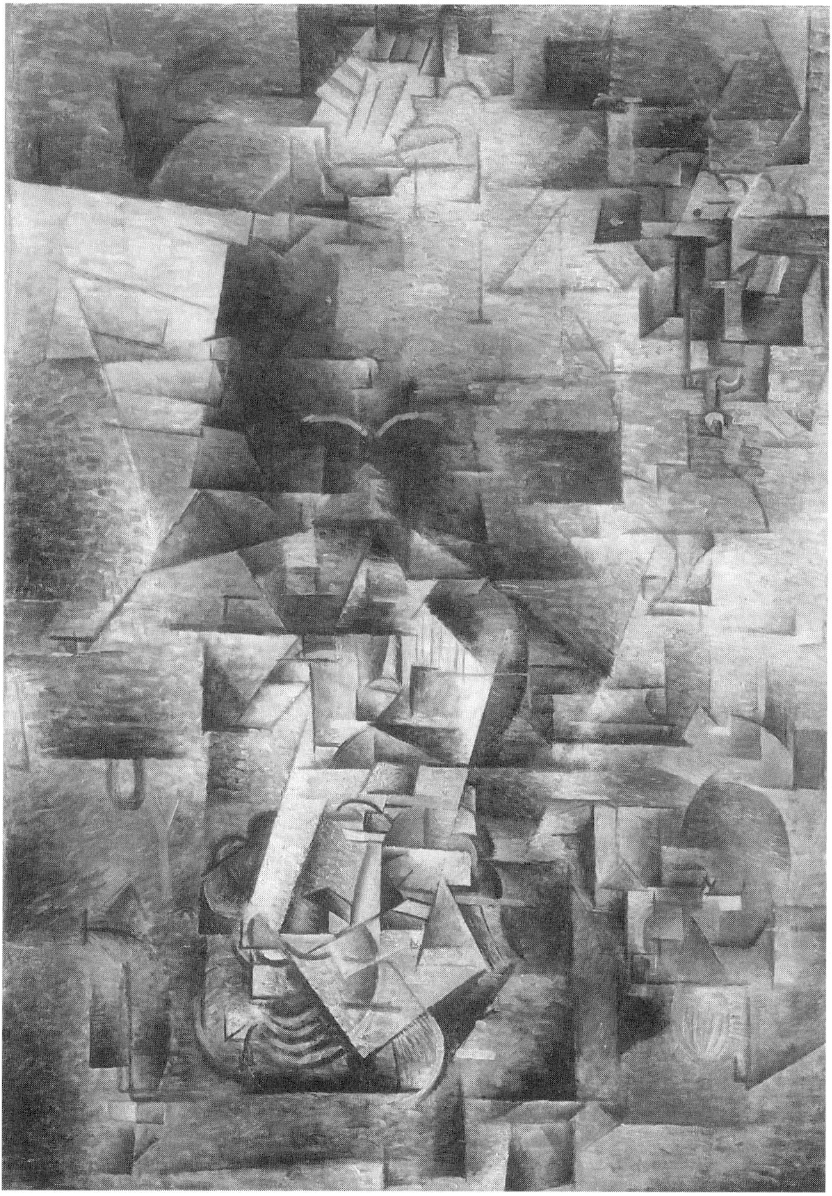

FIGURE 16.1. Pablo Picasso, Spanish, 1881–1973. ***Daniel-Henry Kahnweiler***, 1910. Oil on canvas, 101.1 × 73.3 cm. Gift of Mrs. Gilbert W. Chapman in memory of Charles B. Goodspeed, 1948.561. Reproduction, The Art Institute of Chicago. COURTESY OF ©2004 ESTATE OF PABLO PICASSO/ARTISTS RIGHTS SOCIETY (ARS), NEW YORK.

world. That's part of the point. The changing world is itself neither good nor bad. It just *is*, although we humans sometimes find it exciting. The author of a book about Cubism says something similar about its lack of human warmth. "It was the deliberate disregard of the more sensuous and immediately appealing aspects of painting, and the temporary dismissal of all human and associational values ... that allowed the Cubists to distort and dislocate figures and objects to a degree hitherto unknown."[2]

The Cubists "sought to represent spatial elements ... by alternating points of view. Only when this is properly understood do the changing contours, the displaced eyes, etc ... become intelligible, for in reality [the viewer] is observing different profiles, whose purpose is to achieve a timelessly valid aspect instead of the fortuitous quality of a single view."[3] A *timelessly valid aspect* sounds great, doesn't it? That is just what scientists wish they could claim. (What usually prevents that indulgence is science's dogged devotion to finding flaws in its work.) Likewise, *avoiding a single perspective* is something scientists do to achieve objectivity and to minimize error. Indulge here a reduction of Cubism to the basic impulse behind it: Cubist painting can be seen as an attempt to get on board the fast-moving choo-choo train of the ever-expanding panorama of science.

A few decades before Cubism, visible brushstrokes startled us for a while before they faded into the picture. Now the lines of mathematics plus multiple viewpoints overwhelm the human at the center of Kahnweiler's portrait. Poor guy, he hasn't got a body any more. And poor subjective-minded us, our own bodies have been taken over by cause-and-effect medical science and technology. By the start of the twentieth century, objective knowledge had long since shown us how blood circulates to nourish organs, but it didn't seek a soul or spirit behind the flesh and bones. Likewise, progress now allows twenty-first-century science to look into the human genome for roots of our brainy bodies but not our minds/spirits/souls.

Turn your attention now to Picasso's *Les Demoiselles d'Avignon* (The Young Ladies of Avignon), 1907 (Figure 16–2), an early, somewhat Cubist picture. To say the least, it drew a lot of attention back then. That's because this picture "marks a radical, and at first sight frightening, departure.... We have a demonically distorted world.... There is no source of light, just as there is no perspective vanishing point.... The figures and objects have also been violently deformed, cubed, facetted, stylized...."[4]

In prominent part the picture was successful because, like other Cubist works, it represented the very *idea of innovation*. Modernization itself—useful, successful experimentation—was an admired aspect of the new early-twentieth-century world that was appearing as a result of science

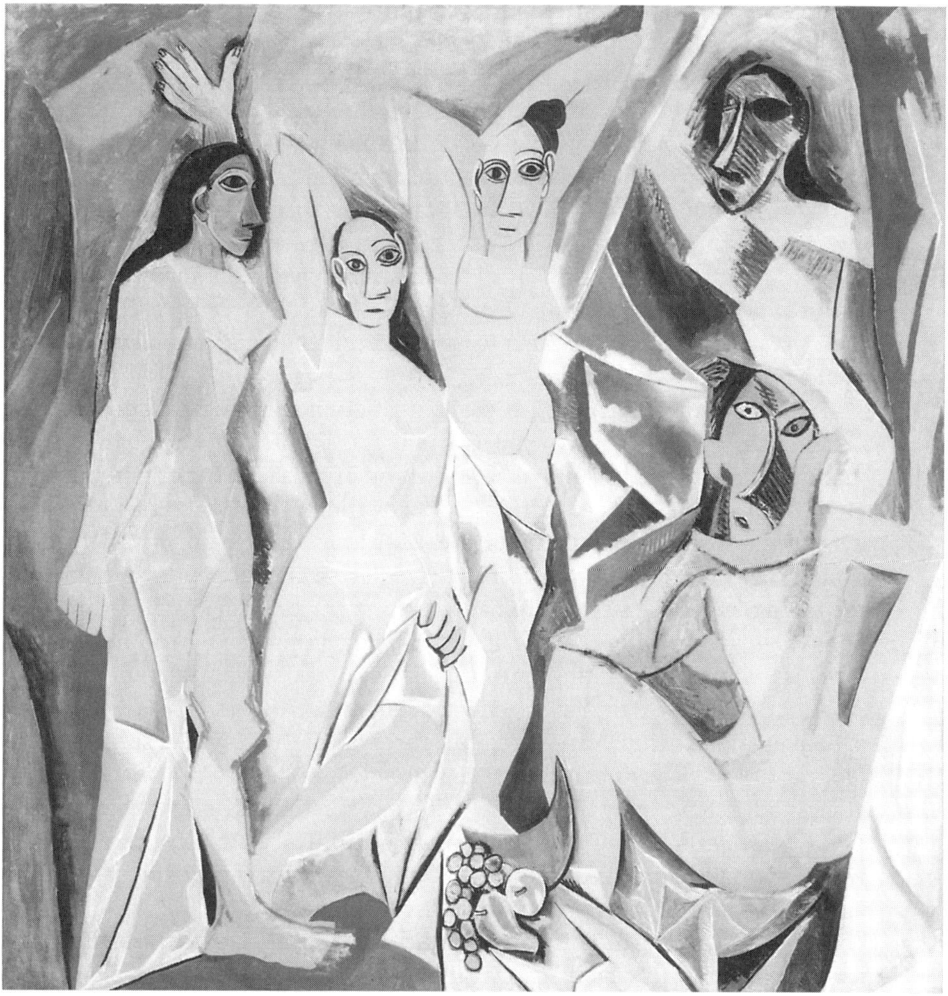

FIGURE 16.2. **Pablo Picasso (1881–1973).** ***Les Demoiselles d'Avignon,*** **1907. Oil on canvas, 8' × 7' 8". Acquired through the Lille P. Bliss Bequest. (333.1939). The Museum of Modern Art, New York, NY, U.S.A. COURTESY OF ©2004 ESTATE OF PABLO PICASSO/ARTISTS RIGHTS SOCIETY (ARS), NEW YORK. DIGITAL IMAGE ©THE MUSEUM OF MODERN ART/LICENSED BY SCALA/ART RESOURCE, NY.**

and technology. In the first decade of the century, the Wright brothers made their first flight in a powered airplane, and the first taxis appeared, as did neon lights and even fountain pens. More weighty developments included Einstein's establishing of the equivalence of mass and energy and Minkowski's formulation of a four-dimensional geometry. Too late by a year to have influenced Picasso's portrait of the young ladies, but not too late for the real sex workers who modeled for the picture, was a cure for syphilis discovered by Paul Ehrlich.

If we look into the picture of the Avignon ladies, there is one thing we can say for sure. They are not best described as gorgeous. But then, we are confronted with the fact that they are common house-bound hookers. Is that, in some perverse way, why Picasso's picture is totally lacking in any attractive erotic quality? Was the twenty-six-years-young Pablo—when he asked the ladies to stop all that wanton wiggling of their hips and hold still for a group portrait a century ago—was he morally resentful of their presumably easy life? That was most unlikely for Picasso, who was then a young goat.

The painting was made, in part, in reaction to Matisse's *The Joy of Life*. Instead of "an outdoor scene of happiness, [Picasso's] would be a dramatic brothel scene."[5] That makes some kind of sense. Like it or not, there is—for most men — some truth in the correlation of easily accessible sexual objects with the happiness of life. However, there is in this picture something decidedly odd about the idea of such a connection. Just look at the young ladies themselves. What's striking for presumably hot young hookers is that there is nothing at all titillating about them. Another odd thing is that they look almost as much alike as objects off an assembly line. Although the two on the right (with the somewhat African mask features) are alike only in being horse-faced, the three on the left (with eyes blankly looking at the viewer and noses seen sideways) might be a preview of the clones that science may deliver in the twenty-first century.

*Les Demoiselles d'Avignon* was said to be influenced by Cézanne's pictures of female bathers. Although totally unlike Picasso's painting in style, Cézanne's *Large Bathers II* (Figure 13–2), for example, does take a remote and non-erotic view of female flesh. With *Les Demoiselles,* Picasso shows a similar interest in what can be called the *objective* side of men's view of women's bodies at the start of the twentieth century. The painting provides an objective *reaction* to the subjectivity that dominates *The Joy of Life* of Matisse.

Although the picture's significance lies far more in its emphatic novelty than in its calm objectivity about sex, Picasso does show the ladies as mere commodities, as sex objects that can be hired. They all can serve the same purpose. What matters mainly is that all of these women are young, unclothed and receptive to sexual commerce. So the ladies are seen with shelves or protrusions on their chests, not breasts. Although they are mostly bare, we see no sign — not even pubic hair — of what here amount to the important working parts of their bodies, the sexual organs in their crotches.

(Note, in passing, that pubic hair is not always found to be erotic. The author of *The Anatomy of Disgust* tells us that when John Ruskin, who knew little about real women, discovered that the one he had wed did not look like the hairless Greek statues of bare women, he was filled "with such disgust that he was unable to consummate his marriage...."[6])

To be sure, *Les Demoiselles* is no more the work of an artist-scientist than Christian Science (created half a century earlier to teach the unreality of pain and death) was the work of a Christian scientist. Yet, in its insistent freshness and its detached objectivity about the young women's work, the picture suggests a reflection of ever-changing science where everything is interesting as an object.

To go back to Picasso being a prude about sex, such an idea is no more than a bum joke. If you have the slightest doubt about that, a glance at his much later non-Cubist *Vollard Suite 85: Bacchic scene with Minotaur*, 1933 (Figure 16–3), will dissipate it. The poses, the champagne glasses in the hands of the man and the minotaur, and the smile on the man's face (minotaurs can't smile) suggest simple pleasure, certainly not the painter's prudery. However, what to make of the two young women in this picture is not clear. It could be that the unsmiling models are amateur providers of pleasure or, like the generation-earlier young ladies of Avignon, professionals. And the males, having paid their last bucks for the champagne after engaging the women, can't afford to share the bubbly intoxicant. It could also be that Picasso paints what was common in his time and not absent in ours, a couple of piggish male chauvinists who won't share even their champagne with their girlfriends. As for Pablo himself, his connection "with women was problematic.... Women close to [him] were often destroyed by the nature of the relationship. The charge of misogyny is often laid at his door."[7]

We should take comfort where we can, however. Note that Picasso's minotaur, in particular, is vastly advanced socially from the first of his line, the original Minotaur. That first bull-headed man-bodied creature was the offspring born of the union of Pasiphae (wife of the mythical King Minos of Crete) and—you are not surprised—a gorgeous white bull of whom she was enamored.[8] For that first minotaur, being even a low-down woman-hater (let alone being advanced as high as a lout) was far beyond his dreams of civility. The first Minotaur was devoted to devouring the Athenian young people that showed up in the labyrinth where he hung out.

To conclude our look at Picasso, we will—only for now—jump far ahead in the timing of this story of art's modern mirror. Picasso painted what may be his most widely known picture, *Guernica* (Figure 16–4), in

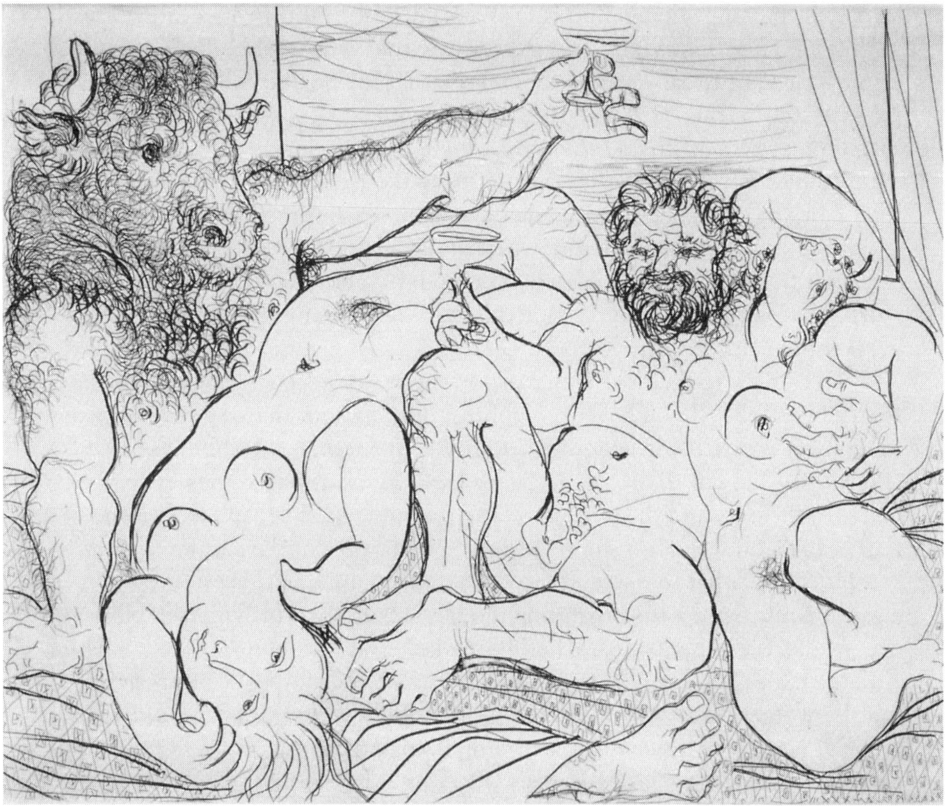

FIGURE 16.3. Pablo Picasso (1881–1973). ***Vollard Suite 85: Bacchic Scene with Minotaur.*** **May 18, 1933. Etching on copper. Inv MP 1982 144. Photograph RMN: B. Hatala. Musée Picasso, Paris, France. ©2004 Estate of Pablo Picasso/Artists Rights Society (ARS), New York. PHOTOGRAPH COURTESY OF RÉUNION DES MUSÉES NATIONAUX/ART RESOURCE, NY.**

1937, three decades later than the *Kahnweiler* and *Les Demoiselles*. Certainly the town of Guernica (whose Spanish Civil War bombing is commemorated) is known around the world today because of the painting. The powerful picture's content is clearly the pain and anguish of people during the bombing. A glance at the mural shows their sorrow and suffering. However, it does not take a close inspection to suggest there is something strange about the pictured people. Apart from their agony, they look to be almost cartoon folk. Their bodies are not only grossly distorted but, for the most part, flat — nothing at all like what people looked like in art before modern times.

How could such caricatures of human beings move us so? It is likely

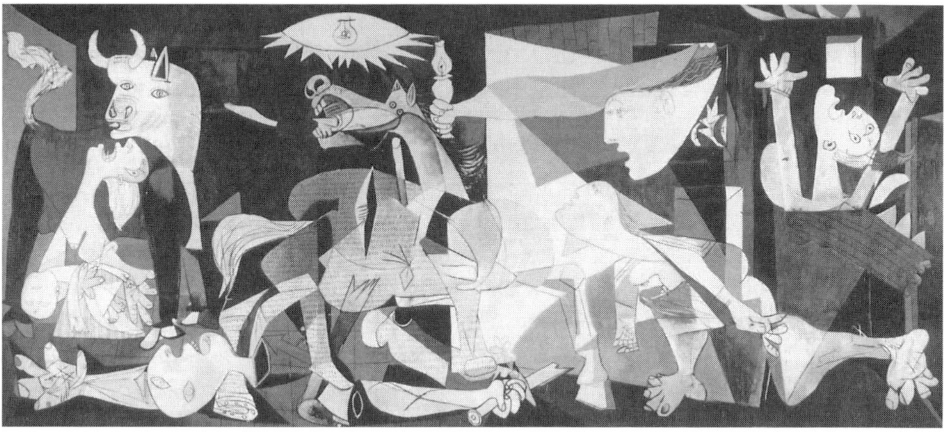

FIGURE 16.4. **Pablo Picasso (1881–1973). *Guernica*. Paris, June 4, 1937. Oil on canvas, 349.3 × 776.6 cm. Museo Nacional Centro de Arte Reina Sofia, Madrid, Spain. ©2004 Estate of Pablo Picasso/Artists Rights Society (ARS), New York.** PHOTOGRAPH COURTESY OF ERICH LESSING/ART RESOURCE, NY.

that *Guernica* "works" in large part because of the contrast between the intensity of those people's feelings and the inadequacy of their feeble bodies. We saw what began to happen to bodies in the *Demoiselles* and the *Kahnweiler*. Picasso's pictured bodies were then already starting to lose their depth and their reference to real bodies. "Beginning in 1912, [with the Cubist] work of Picasso and Bracque ... the pictorial illusion takes place upon the physical reality of an opaque surface rather than behind the illusion of a transparent plane."[9] Pictures that moved from the depths of space to its surface led to the mostly two-dimensional bodies that we see in the much later *Guernica*.

We can see that, over the decades, human bodies were increasingly becoming not only things but *mere* things, with that "nothing much" quality represented by their loss of a third dimension. This was happening even as the minds within those bodies remained the means whereby people endured pain and suffering.

No one would dispute that we had learned much about the nature of animal and human bodies in the first third of the twentieth century. Although science brought us that knowledge, it did not suggest that bodies were of reduced significance. Yet an intuition to that effect began to present itself in our minds. Why would that have been? In part because the work of Darwin and other scientists increasingly bore the weight that once was carried by biblical accounts of our origins, at least of our origins as bodies. Think of Michelangelo's Sistine Chapel ceiling fresco celebrating the moment when God's touch brings Adam's body to life. That kind of grandeur is not to be found in bodies whose origins lie in countless modifications resulting from the indifferent touch of cosmic rays that mutated DNA molecules.

Uncertainty about the likelihood that bodies were the non-reducible creations of God reduced their significance to us. Yet our origins as *minds* remain unaccounted for by science to this day. It is easier to identify our unseen minds with our also unseen spirits, and thus with an unseen spiritual almighty God in Heaven above. *Guernica*'s people suffer — even though their bodies don't amount to much — because, inside those utterly flimsy bodies, they are still spirited subjective entities who hope God will spare them more anguish.

Within a few decades, art would depict new horrors, not only physical ones (although there were plenty of those) but conceptual frights that take the form of Pop-art people with minds and spirits that are as lacking in depth as are the bodies in *Guernica*. Hold on to your paper hats, the party has barely begun.

# *Chapter 17*

# "But Is It Art?": Marcel Duchamp

To get some understanding of the significance of Marcel Duchamp's creations, we will explore his breakthrough in what was acceptable as a work of art. Irreverence was a large part of Duchamp's attitude toward the art of his time. So it is fitting to copy his attitude by calling the big fuss about one of his works a good-sized tempest in a pee pot. Well, not a storm regarding an old-fashioned pee pot, but about a modern one, a urinal he offered to the art world as art.

Note first an age-old view about art. "Leonardo da Vinci thought art should be 'a thing of the mind.' And [as it happens] Duchamp thought so too."[1] Despite that, his transformation of a urinal into a work of art raised doubt about whether much in the way of *mind* was needed for art.

What is art, and what isn't? That was a subject that, with Duchamp's work, demanded attention. Philosopher Arthur Danto puts the issue more fully, roughly like this: An art-object might look just like something that is merely a plain object, so just what is the difference between the two in such a pair?[2]

As an example of what he's talking about, Danto uses Marcel Duchamp's *Fountain*, 1917, replica 1964 (Figure 17-1), which "was nothing but a urinal of that period disconnected from the plumbing ... and turned on its back like an immobile turtle...."[3] The problem emerged when *Fountain* was publicly displayed as a work of *art*.

What *is* the difference between a urinal as a work of art, and a urinal as a modern pee pot? Glancing through his book didn't reveal Danto's answer. So here's a useful explanation, one which will then be justified.

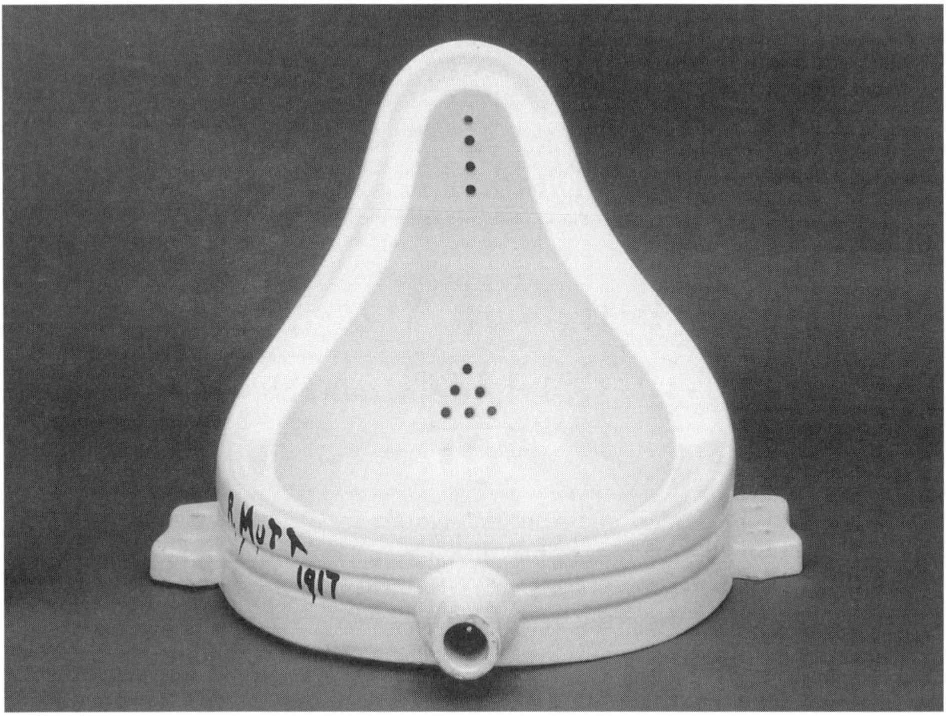


Figure 17.1. **Marcel Duchamp (1887–1968). *Fountain*, 1917, replica 1964. Porcelain, unconfirmed: 360 × 480 × 610 mm. Purchased with assistance from the Friends of the Tate Gallery 1999. Tate Gallery, London, Great Britain. ©2004 Artists Rights Society (ARS), New York/ADAGP, Paris/Succession Marcel Duchamp. Photograph ©Tate Gallery, London/Art Resource, NY.**

First (and by itself not very deep), a work of art is an object that much of the art community *regards* as a work of art. Second, for a member of that community to regard something as an art-object, that person must have a *subjective mental concept* attached to it. As a member of that group, you have something in the way of an *idea* about it in your head. Such an often inarticulate idea might not be *expressed* fully or well by putting it in words: A certain work represents or is about "such-and-such."

To justify the second part of that answer (that art has a subjective mental component), let's start by considering, not an object-as-art and a physically identical object-in-itself, but a different pair of things. Look at a *gesture* (in the sense of someone making a deliberate significant movement) and a physically identical *behavioral-act* (an act that has no thought behind it). What's the difference between these two?

For an outside observer, there certainly are no perceived differences that distinguish one from the other. So what is the distinction? A gesture is such precisely because it *embodies a* (subjective) *concept* that is absent in the person performing the similar-looking behavioral act. Let's say you briefly close an eye in front of someone. Only if you have a related "something" in mind when you do so is that a gesture in the form of a *wink*. But if your mind was elsewhere, what you did with your eye was not a gesture; it was merely a behavioral act, a *blink*. (The observer may later use other clues to "get with it" and realize that the brief eye-closing was a wink, not a blink.)

The same is true of the artwork; there is a concept behind it. For sure, once a work such as the modernized pee pot carries the label of art, its new significance is not easily stated or agreed upon. Nonetheless, various versions of it are really there, in viewers' subjective minds. For a related instance, if several of us watch you hold your thumb toward your nose and your forefinger away from it, there can be agreement that you have gestured, but disagreement about the idea behind your gesture: Were you thumbing your nose, or were you pointing a "gun?" (A useful understanding of the import of Duchamp's artful pee pot is provided below.)

In *The Transfiguration of the Commonplace*, Danto makes a somewhat similar point. "The difference between a basic action and a mere bodily movement is paralleled in many ways by the difference between an artwork and a mere thing...."[4] Although Danto does not spell it out in terms of *consciousness* or *awareness*, a "basic action" differs from a "bodily movement" in that the former has a *subjective mental* component that is lacking in the latter.

As we know, there is a problem with inquiring about subjective mental activity: From the outside, no one can be sure about what's going on in some other person's mind. This is part of the old subjective-mind versus objective-body problem. The many meanderings of other (subjective) minds are, in themselves, an ongoing mystery because — although frequently inferred — they are never directly available to outsiders.

Never mind that now. It's time, perhaps past time, to seek the import of Marcel Duchamp's *Fountain*. But let's not get hard-headed or literal-minded, like "George Dickie, the institutional theoretician of art [who asks,] 'Why cannot the ordinary qualities of *Fountain* — its gleaming white surface, the depth revealed when it reflects images or surrounding objects, its pleasing oval shape — be appreciated?'"[5] They can indeed be satisfying, George, but that's not the point.

First, we should acknowledge the likelihood that Duchamp was not only making art, but was also thumbing his nose or pointing a mock gun at the art establishment. He thought that "much of the avant-garde art ...

was in fact rubbish, of less significance ... than many objects of everyday use."[6] Despite that, the urinal/*Fountain* object was (eventually) accepted by much of the *community*, those who create, display, write about, and look at art. So let's claim membership in that community's majority, and treat it as art. Let's also continue to follow philosopher Susanne Langer in treating art as basically "a transformation of subjective ... realities into objective ... forms."[7]

The significance of *Fountain* is—precisely, if perversely—this: It initially appears to be an object that has *little* significance. This may seem a paradox but it really isn't one. The import of *Fountain* is that *meaning* (which in itself is associated with ideas in our minds) was becoming of less importance in people's thinking at that time. At the same time, *information* was impressing us more and more. Here's a vulgar and trivial example: "This piece of porcelain is a thing that men piss in." With his urinal, Duchamp is slightly supporting but mostly challenging the belief that subjective mind is significant. He supports it a little by (implicitly) asserting that the object is art. He challenges it a lot by making that claim about an artifact that—as a urinal—doesn't have a good deal of significance. That is why there was such a tizzy in thinking about *Fountain*. Is this thing to be or not to be regarded as an artwork?

By presenting the urinal in 1917 as a *Fountain*, Duchamp showed the *spirit of the times*. Early in the twentieth century, people in Western society began to get disinterested in the idea of significance. That's because important meaning itself began to thin out from areas like religion, spirit and soul, where once it was strongly established. At the same time, new meaning of a less important and more pedestrian sort was showing itself with all the information, technology and objects flowing from new objective knowledge.

For the decline of old meaning, we need look only for a moment at religion. It used to be for almost everyone (and indeed still is for many) that religion was fully loaded with import and significance. Listen to this portion of the Second Commandment. "...I, Yahweh your God, am a jealous God and I punish the father's fault in the sons, the grandsons, and the great-grandsons of those who hate me; but I show kindness to thousands of those who love me and keep my commandments."[8]

It hardly need saying that the significance of new things such as newspaper comic strips, the parachute, and the automobile assembly line (all of which came into existence in the decade before a urinal became the *Fountain*) doesn't begin to bear comparison with the import of God's Commandments. A fair comparison, however, is that of the concept of God in the Bible with the idea of God in the eighteenth and nineteenth centuries.

Then, with the arrival of what was called Deism, God became not much more than a First Cause. This new version of God is not much involved with the day-to-day running of the physical universe. Nor is it clear that the Deist God is either punitive or kind to us as living individuals in His huge universe.

If you believe in the God of the Second Commandment, He provides an abundance of (subjective) significance in your life. His successor for some, the Deist First Cause, doesn't occupy as much mental space, although for those that accept Him, He is better than nothing at filling lives with meaning. (No things or spirits fill the empty supernatural realm in the minds of those who don't believe a real one exists.)

The new Deist God came with the expansion of science into vacant mental territory once occupied by the earlier idea of God. Astonishingly, science got by largely without the idea of subjectivity. Its wildly successful method of accounting for things was, in part, to concentrate on cause-and-effect explanations that were useful. The information that resulted began to explain inanimate objects, living animals, and much about humans, immortal soul or not.

What also began to exist and to work nicely was—for a pisspot-pertinent example—the plumbing. Running water—what a luxury—began to show up increasingly in houses. With the arrival of sewer pipes, the under-the-bed piddle pot, once the staple of every home, began to disappear. It was replaced by the flush toilet—and in public places (in the patriarchal society that found easing the needs of men more important than doing the same for women), by the urinal. In a light-hearted sense, we could say that the pleasures of plumbing provided—hardly compensation but—something that filled a corner of the new void due to the diminution of the likelihood that God was watching our every move.

Let us return to and draw some conclusions about Marcel Duchamp. The urinal that he named *Fountain* was accepted as a work of art because it was like a complex gesture that embodied the new importance of human-made objects and the new reduced significance of subjective concepts. The gleaming white ex-urinal reflected the confusion and uncertainty about the importance of meaning. It might even have served to convey the bare beginnings of a certain queasy uncertainty about the reality of our own subjective nature: Are we as *subjective* entities that are nowhere to be seen in the *objective* world—are we as minds, spirits and souls, relative to our bodies that include brains, all that important? As we will see in chapters to come, some scientists are now suggesting that there is little or no import to our subjective nature.

Although some in his era may have been interested in the importance of objects, Duchamp himself might have been more interested in having fun with seemingly meaningless art objects. "In 1913 he mounted a bicycle wheel on a kitchen stool and called it a 'distraction.' That winter he ... worked on *Three Standard Stoppages*, an assemblage of three-metre lengths of string glued to canvas on ruler-like strips of wood and fitted into a box.... In 1914 he issued the first 'readymade' when he bought a bottlerack and inscribed it as a work of his own, subverting notions of authenticity and invention...."[9]

In another half a century, Duchamp's paradoxical pisspot would gain some company. For example, Andy Warhol would badger us with objects such as Brillo boxes and soup cans that carried on the assault on subjective significance. One of the art movements that co-existed with the Pop art of Warhol and others was called "Conceptual art." In terms of seeing meaning as *important*, Conceptual Art was at the opposite end of the seesaw from Duchamp's *Fountain*. Sol LeWitt and other Conceptual artists in the 1960s insisted that the *idea* itself was the important aspect of an artwork, and that the actual object was secondary.

All of that raises one last issue that should be addressed: Is it a *contradiction* to suggest that at the same time in the 1960s, art (such as Warhol's) was about worldly objects, and art (such as LeWitt's) was about mental concepts? Well, no, not at all. Art is not like rocket science; art reflects how we humans make (sometimes contradictory) meanings. As for *meaning*, here is a one-sentence summary from anthropologist Clifford Geertz: "The imposition of meaning on life is the major end and primary condition of human existence."[10]

# PART V

# Looking Outside and Within: Early Mid-Century

***Chronology of Technological and Scientific Developments***

1930: Yellow fever vaccine discovered

1931: Heavy hydrogen (deuterium) discovered

1935: First sulfa drug for treating streptococcal infections

1939: Possibility of splitting the atom is demonstrated

1939: FM invented

1940: First successful helicopter flight in the U.S.A.

1941: Plutonium discovered

1942: First computer developed in U.S.A.

1943: Penicillin used successfully on chronic diseases

1945: First atomic bomb detonated

1947: Scientists invent transistor

1948: Mount Palomar 200-inch telescope dedicated

## *Chapter 18*

# Turning Away from a Darkening World

Remember when there was an Enlightenment? That was part of an upbeat new emphasis on rational thought and on the expansion of science, remember? Indeed, it is hard to recollect our once-upon-a-time hopes of casting light into all the dark corners where people had for so long cowered. Well, you needn't be told that those hopes were transformed, if not deformed, as we approached the middle of the twentieth century. The long-gone Enlightenment had been succeeded — not by a new dark age, but by a mottled epoch, a crazy quilt of new dark patches together with new bright ones.

Part of the darkness in the new world was the 1914–18 "War to End All Wars" that killed ten million with both old-fashioned and new weapons. The war's name had to be updated to "World War I" in the 1940s, thanks to World War II, which made no optimistic promises and which killed far more soldiers on battlefields and far more civilians in cities. By the middle of the century, World War II had been succeeded by the Cold War, which knew better than to promise anything. It only hoped to try to keep us from what was clearly possible: the madness of "Mutual Assured Destruction" with nuclear weapons.

Another part of the dark new objective knowledge brought about dictatorships with the machinery to control hundreds of millions and to kill tens of millions. Yes, there was nothing new about powerful people who could control everyone they could reach and kill those who didn't fit in. But the old ones were small fish compared with the giant sharks and tentacled octopuses with the technology that helped control billions.

Knowledge and artifacts to control and kill huge numbers of people — that was one thing. What about medical science that could and did save millions with drugs, vaccines, and antibiotics? That was no killing field, that was new enlightenment, yes? It surely was. Just as surely, it fueled a population explosion that made it in our interests to clear the earth of many of the plants and animals whose space we wanted for ourselves.

As social scientist Max Weber pointed out long ago, modern people "live in a 'disenchanted' world. The process of disenchantment began with the rise of civilization, but it ... has been greatly accelerated by the rise of modern science and technology. While science gives us all kinds of knowledge and power, it is 'meaningless' in that it does not answer the most important question: 'What shall we do and how shall we live?' Religion had answered this question, but it has lost its authority; popular religion gives most people no real certainty for their living purposes.... Hence, Weber observed, both life and death may become meaningless."[1]

Although Weber was mostly right about science, and right to some extent about religion, he did not foresee the comeback that popular religion would make. Between 1910 and 1912, millions of copies of a publication called *The Fundamentals* were sent to religious workers. That publication "launched a movement [of those] who feared the collapse of ... a 'Protestant consensus' in America.... Millennialists and biblical literalists agreed that World War I was a God-sent event to urge the world to do penance for its sins.... The most notorious aspect of American fundamentalism was its crusade against Darwin's theory of evolution.... The Bible Crusaders of America, founded in 1925, ... published its views in the *Crusaders' Champion*. The magazine embodied the fusion of fundamentalism and politics for the sake of saving America from the evils of modernism and evolution."[2]

By the 1950s, this new revival of religion was backsliding a bit. It needed to assert some claims to objective knowledge. "The new fundamentalists attempted to overcome [their] old image ... as an anti-intellectual and anti-scientific movement by arguing that [their truths] were in harmony with reason...."[3] However, fundamentalism still echoed its earlier teachings of the inspiration and inerrancy of the Bible, its millennial prophecy about the Second Coming of Christ, and its belief in the salvation it brings to be "born again" into a new life of love. So, concessions to modernity or not, fundamentalism was successful for many because it re-enchanted the "endarkened" world with its strong appeal to subjective belief that Jesus can save us from our worst selves.

Apart from American fundamentalism, "the flagship of the reaction to science was Oswald Spengler's widely read 'Decline of the West' (1918), in which nonrational, mystical ideas were touted as the antidote to the scientific subversion of civilization and nature."[4] At another level, the humanities, once the big brother of the science culture, became embittered about its expanding rival. "One of the roots of the fairly prevalent antagonism of academic humanists to science [may be] an unconscious resentment of the emergence of a high-profile, worldwide community that sometimes implicitly or explicitly claims to be involved in a pursuit that is more important than the humanities.... A distinguished historian like David Thomson writes that in the twentieth century, 'the scientist ... came to dominate the whole sphere of creative endeavor.'"[5]

At about that time, American art, which had been representing the hard times of the 1930s with what was outward-looking Realism, flipped over. It started representing the new turn away from knowledge about a world that had become ever more frightening, grisly and grating. Art turned away from the darkening outer world and toward — with Abstract Expressionism — our mysterious inner realms. "The chief theorist of Abstract Impressionism [Harold Rosenberg] describes the style as a 'conversion phenomenon.' He goes so far as to call it 'essentially a religious movement.'"[6]

This new style was an extension of Expressionism, which refers mainly to the work between 1905 and 1920 by German artists. "If indeed Expressionism means anything it means the use of art to transmit personal experience.... Instead of reinforcing the false values of a materialistic society, art thus used would help people to recognize their own spiritual worlds."[7] "It seems natural to suppose that [what had happened to] the substantial and solid world of things would induce many painters to turn to man's inner impulses and psychic experiences.... But in this field there are no really visible things to reproduce or to define.... The result would be a nonrepresentational painting which would so develop the pictorial means ... as to make them capable of rendering man's inwardness. Such a type of painting ... made its appearance with Abstract Art...."[8]

One of the problems with an art that turned so strongly inward is that it may be difficult to say what a particular picture is about. That is something like the outlandish religionist talk called *speaking in tongues*— glossolalia. For insiders in the religious community, the particular "words" spoken by someone in that way matter little. Rather, the entire activity of so speaking is regarded by Born-Againers as coming from the Holy Spirit and providing evidence that the speaker has been reborn. Many of those outside that community, however, don't know what it means when someone speaks in tongues. To them, it may sound like mumbo-jumbo.

The art community classifies certain inward-facing pictures as

Abstract Expressionism. But what does a particular picture mean? Here, the artist himself sometimes comes to the rescue by virtue of the words of a defining title. For many viewers, no title is needed to feel a picture's impact or high significance. For others, even the title provides little help. They can read in a given picture no more than a mumbling jumble.

We will look first at two artists, Edward Hopper and Francis Bacon, who stayed with the harsh real world. Then, with Abstract Expressionists Mark Rothko, Willem de Kooning, Lee Krasner, and Jackson Pollock, we will turn inward to their (for some, magically evocative, for others, little more than paint) subjective art.

## *Chapter 19*

# Alone in the Crowd: Edward Hopper

Early in the twentieth century, a sociologist said the following about living in modern cities: The large-city dweller, "who is rootless and traditionless compared to his country ancestors, places his faith in economic and scientific values. Thus alienated by [this] 'commodification' [of life], ... modern urbanites maintain relationships [and make judgments] by observable facts and quantitative measures."[1] Those words, although rather extreme, give us a point of view for the paintings of Edward Hopper.

Before we explore some of this artist's work, let's remind ourselves how modern large cities are related to expanding science. Apart from the networks of sewers, water pipes, streets, electricity wires, and related city necessities, there is the basic fact of large cities themselves. They exist because we became so much more efficient at the primary task of providing food for ourselves. That has enabled rapid population increase, reduced the need for more people to produce our food, and added the need for people to make new things in those expanding cities. All of that flows from the fountain of science and technology.

The isolation of the people in Hopper's pictures, even of each partner in a couple who share their lives, is a major theme for this artist. His *Excursion into Philosophy*, 1959, shows a man sitting in thought on the edge of a bed with a book open beside him. Behind him on the bed lies a near-naked woman, facing the wall away from the man. The artist makes the picture strongly project a sense of the man's isolation, and it suggests that of the woman as well.

Hopper's wife, Jo, made the following note about *Excursion into Philosophy*: "The open book is Plato, reread too late." Hopper's biographer comments on her note: "Since Plato praised the realm of ideas as the ultimate form of reality and relegated physical manifestations of them to a lower realm, the man in the painting seems to be questioning ... the idea of beauty versus the presence of the voluptuous female on the bed beside him."[2]

Perhaps Jo Hopper's remark was suggesting something rather more specific. If the man had reread Plato earlier he would not have descended from the world of ideas to seek satisfaction where he would not find it—in the physical world of the flesh. He wouldn't be sitting forlornly with a semi-nude woman reclining behind him. Peering pruriently into the picture would tell us the bed still has a cover on it, and it isn't the least bit rumpled. So chances are they have not been playing the interactive game of physical love. It is likely that, even though they are hardly old, they just don't do that any more. Not only do they not pleasure each other physically, but it's doubtful that he was offering to gratify her mind by reading to her. They may still be together, but they cannot give each other much. Maybe he sits there wondering why they are alone even though they live together. As for Plato and his realm of ideas, the man has put the book down. He can't find companionship there either.

Talking about sex brings us to Hopper's *Study for Girlie Show*, 1941. The name "Girlie Show," its equivalent, "striptease," and the nature of the show itself are now out of date. Hopper shows a flashy, fleshy woman who does nothing fancy by current standards; she just exhibits her body on stage in front of a mostly male audience (not shown in the picture). The idea behind the picture, however, is more than merely still with us; it is behind the huge new business of selling erotica and pornography to men.

The picture shows that twentieth-century men—lonely men in particular—could purchase something new and different. Their predecessors in previous centuries could and did purchase sexual activity as a commodity. Twentieth-century men also purchased the mere sight of uncovered female flesh as a sex commodity.

The essence of the mature girlie's performance is a piece-by-piece removal of her clothing. That accounts for the *strip* part of the synonym. What about the *tease*? What does that mean? As she does indeed deliver the sight of her body bared, she is certainly no teaser about that. Yet the name "striptease" for the performance does seem right. That's because the tease is really about something beneath the stripping. Although a bare moment's thought suggests the tease is simply about displaying herself as an *unavailable* object of desire, that answer is rather murky. It doesn't tell us why it is that stripping off her garments makes her desirable to her male audience. After all, those lonely men surely had no doubts, before she stripped, about just what was beneath her clothes. So there is more here than meets the eye. It is not simply that she is desired; she is coveted for a hidden reason. The reason is that her baring of her body makes her appear to be a *receptive* female, one who would welcome sexual activity. What excites her audience is a woman who looks like she has stripped for action, sexual action.

That leaves this question: Why is it that a woman who has taken off her clothing is commonly regarded as sexually receptive? The barefaced answer — that's just the way men are — is good enough for most of us most of the time, but it doesn't say much. After all, that's not the way men are in tropical climates. There, the sight of women without much clothing does not, in itself, excite men.

For a more informative answer, start with something from scientists' observations of the behavior of creatures far removed from this city scene — our country cousins, the gorillas. Gorillas *in captivity* have been observed to engage in sexual behavior as follows. When a female, even one who is not in heat (is not sexually receptive) happens to come near a male, the male first undertakes a chest-beating "dance" of sexual excitement. After that, he engages the female in copulation. This male behavior was initially mysterious to scientists. The conduct was understood only when it was realized that, *in the wild*, a female gorilla goes near a male only when she is in estrus (and thus is sexually receptive). A male in captivity understands (wrongly so, dumb brute that he is) the approach of the female as a *gesture* that represents her desire to have sex, rather than as a mere consequence of confined captivity.

As— in most ordinary familiar circumstances— both the female ape's *coming close* and the female human's *body-baring* is associated (in their males' minds, dumb or not) with willingness to engage in sexual activity, the two are equivalent for the ever-on-the-alert attendant males. That's why the stripping activity by the woman in *Study for Girlie Show* suggests to each man in her audience that she is in the mood for physical love.

If that's so, why don't some of those men in her audience rise out of their seats and show they are not dumb brutes by forming a cue for her? They don't do anything like that because minds have parts, parts that often have conflicting goals. With brains and minds that dwarf those of gorillas, the lonesome solitary men watching the girlie (as she shows her most prominent sexual parts) may feel like beating their chests, but they also understand that the stripping is only a tease. One of their twenty-first-century peers gets flushed with sexual blood when a mostly naked woman comes over and opens her legs in a "lap dance" on him. Likewise, men now purchase the sight of a rather different performance, that of two women engaging in sexual interaction. Although that sight might suggest that such women are not interested in male sexual partners, it still excites men, as it invokes the basic ancient notion that the performing women are "in heat." Like the men in the girlie's audience, such men's subjective experience gives them inarticulate carnal knowledge: *she acts like she wants it*, and that thrills them, despite their objective understanding that *she is only acting to earn her living*.

Why do male gorillas and humans behave like they do? The males in both species have something in the way of a built-in *genetic* instinct, horny "rut," that moves them to seek a receptive female. For lower mammal species, a receptive female is one with a built-in genetic instinct for periodic "heat." Although human females retain only a small monthly trace of such *heat*, human males retain a good deal of *rut*. Without trustworthy female genital swellings and aromas implying heat, men rely on (unreliable) *symbols* of female receptivity.

Hopper's girlie also shows how standards for the portrayal of young women were changing in the twentieth century. In recent paintings, women are seldom portrayed as *beautiful*, although they may be shown as sex objects. This, along with the frequent absence of a noble loving heart in pictured people, is among the many results of the profound darkening of our views of ourselves in the twentieth century.

Hopper's stripper, although more coarse-looking than beautiful, is indeed sexy. It is of interest to note that what makes her erotic is not just her nudity, but also her curvaceous relative youth. Before painting the picture, Hopper made a "Drawing for painting *Girlie Show*, 1941," for which his then-middle-aged wife posed. A glance at each shows that the painter not only changed the girlie's facial features from those of his spouse, but also slimmed her waist and lifted her breasts, thereby adding voluptuous youth. The reason that men perceive more erotic quality in younger women is not really established, although there is a scientific-sounding explanation: Younger women can have more babies that can older women, so it must be an adaptation in our genes that makes men find the youthful ones to be sexier. However, evidence from the apes—with whom we share almost all of our DNA—does not support a *simple* it's-all-in-the-genes account. Male apes (in the wild) show sexual interest in any female with private-part swellings or scents that suggest she is hot. Female age, as such, does not matter to ape males. So for human males, what's in the genes gets greatly expanded by ideas in the mind. Indeed, as we all know, both male and female humans are often more occupied with *concepts or symbols* of things than with the basic *things* that get represented.

For the lonely men in her audience, the naked girlie is a symbol, an undulating flag that captures their attention. To them, seeing her represents the freshening churning breeze of a sexually receptive woman. But the sight does no more than represent what it cannot deliver. When those men return to the dead air of their real lives, some may find a real woman to fill their days. Although such a woman may lack the showy acting and the curvy body of the girlie, her real responsiveness to her man will more than compensate for her insufficiency as a visual symbol. As for the girlie herself, we know that as she gains years, she will lose the ability to earn her living by representing an attractive, receptive woman.

There is another way, one that Hopper did not allude to, of looking at the girlie and her male audience. Many people view pornography and its precursors as part of an ongoing *degradation* of women by men. For them, to treat women as sex objects is to debase them, to reduce them to less than fully human beings. They believe men should not indulge their base animal instincts but should resist them; doing so would uplift the men as well as the women. Many modern people also believe that if men lack the courage to avoid such degradation, they can get help, not from science's increasing knowledge of human nature, but rather from religion. They believe that seeking the support of Jesus can help the men.

Critic Arthur Danto considers the professional situation of Edward Hopper: "Hopper and the artists who understood him, and whom he understood, were marginal, and in danger of being pushed off the board altogether [by Abstract Expressionism].... Hopper was included in the Museum of Modern Art's second show, in 1929. [The museum's director] thought Hopper 'the most exciting painter in America' when he gave him a retrospective at that museum in 1933.... By the 1950s, ... art of the sort Hopper exemplified was in danger of being swamped by a modernism narrowly defined in terms of abstraction."[3] Edward Hopper used the old means of portraying parts of the observable scene to exhibit his ideas about how people lived in the "brave new world" that came from science and technology, a world that sometimes brought cheer and goods, and at other times made them cringe with fear. Hopper shows that it also made some people wilt with solitude.

## *Chapter 20*

# When Flesh Trumps Spirit: Francis Bacon

To make sense and symbol of some of Francis Bacon's paintings, it may help to listen first to some words about an interval in the Western world's twentieth-century history. "Never was any period so shaken by such horror, by such dread. Never had the world been so deathly silent. Never had man felt so small. Never had he been so afraid. His misery cries out to heaven ... Art cries out too ... cries out for help, cries out for the spirit...."[1]

Those words do *not* refer to the horrific middle half of the twentieth century. They are about what *preceded* it — the period in Europe from 1905 to 1920 that included the First World War. Near the end of that interval, William Butler Yeats expressed his fear that the breakup of "our scientific ... fact-finding ... civilization"[2] was coming and voiced it powerfully with his poem, "The Second Coming."

That earlier period proved to be only the beginning of what became a continuing avalanche of horrors: Stalin's realm, Hitler's arena, the Second World War, and then new weapons that grew in power and number during the Cold War, and brought the vista of mutual assured destruction of Western civilization.

Although Francis Bacon painted during the dreadful years from the 1940s through the 1980s, he did not portray the terror of that frightening world. In fact, the artist was known for something that seems quite unrelated, his "twisted horrifying figures ... figures as freakish half-human, half-beast.... He [intended that] his deformed portraits would leave 'a trail of the human presence ... as the snail leaves its slime....' [He] called his work 'exhilarated despair.'"[3] However, in some of Bacon's paintings, he connects all of that to the barbarous world around him. Those creations provide a parable about the brutalization suffered by many of his fellow humans who endured living through those savage years.

Among Bacon's works are several paintings of what seems an unearthly creature. He also created a number of pictures of a sitting pope. We will look into one of each of these two kinds of portraits. Separately and together, they render an allegory about the mid-twentieth-century human condition, one that is both hard and hard to deny.

To see a painting of the first kind, turn to *Three Studies for Figures at the Base of a Crucifixion*, ca. 1944, right panel (Figure 20–1), with its long-necked brainless beast. In another portrait of this creature, *Untitled*, 1943–44, there are lines behind the animal suggestive of a cubical frame, but they in no way contain this active creature. If the brute could bend what pass for its legs (only one of them can be seen in Bacon's right-panel study) as it prowls on them, it would be slouching toward our left. Above those tubular "legs" is an oddly hollow area with little room for a stomach and a gut. Yet the beast's mouth is wide open and eager. Eager for what? It is greedy for more, that's what—its appetite is bigger than its belly. It doesn't think about this, nor about anything else. A reasoning brain above its mouth is something it neither has nor needs. What makes it insatiable for more things to devour is simply its basic self-serving nature. As we will see, Bacon's picture tells us that the way of all flesh is to grab all it can get and then dump what it has no use for.

Is the monster a totally *natural* creature? That's clearly not the case—the body-supports it stands on say otherwise. They are not normal "legs"; they look like elongations of the tapered cylinders of human-made tables or peg-legged pirates. That suggests they are *attached artifacts*. Nor is the head that of any ordinary creature. In the *Untitled* version of its portrait, there is a clump of small objects situated just in front of its open mouth. In fact, we can't be sure whether that torrent is entering or leaving the orifice. Perhaps the monster has no other hole—the one with the teeth serves both as a mouth and as a cloaca. One hole or two, the rough beast is not fussy nor is it fastidious.

We can learn more about this monster from David Sylvester's book about the painter. In another painting with this beast at its center, *Figure getting out of a Car*, 1943, the car is "a precisely realistic image [which] is known to have been copied from a photograph of Hitler emerging from a Mercedes at a Nuremburg [*sic*] rally ... [Also, the creature is] a long-necked version, reminiscent of a gargoyle, of the figure in the centre panel of the triptych completed in 1944, *Three Studies for Figures at the Base of*

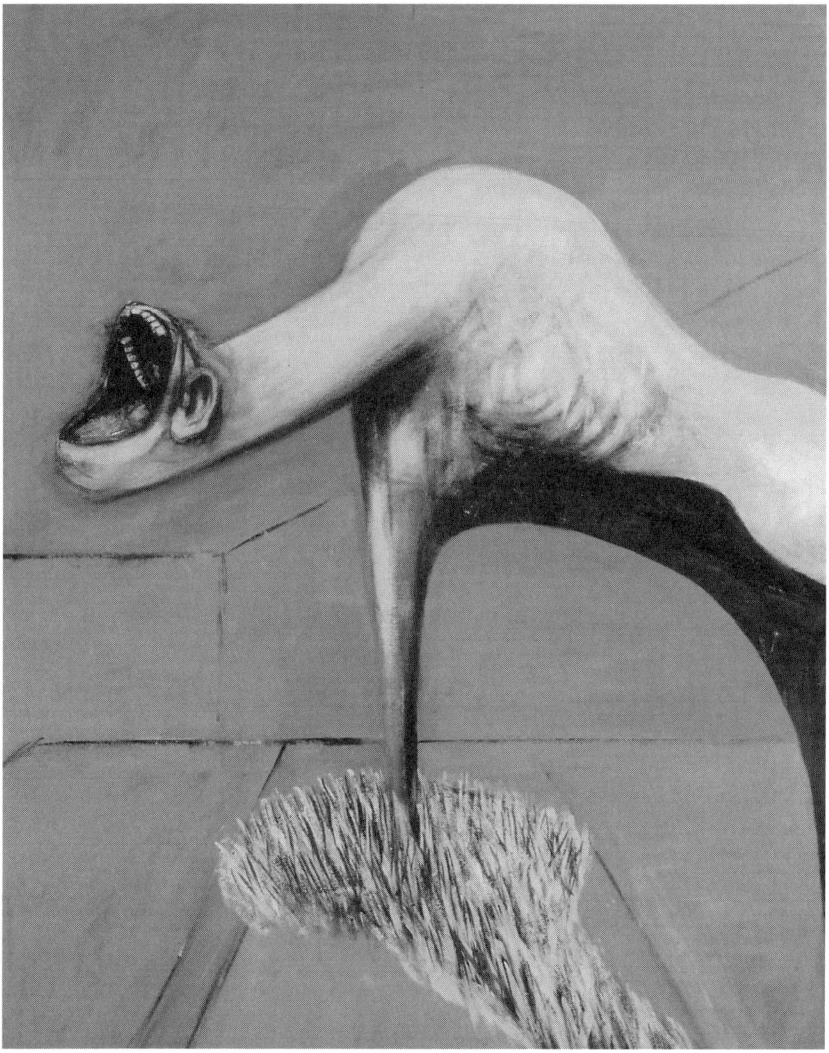

FIGURE 20.1. Francis Bacon (1909–1992). ***Three Studies for Figures at the Base of a Crucifixion***, ca. 1944. Right panel. Tate Gallery, London, Great Britain. PHOTOGRAPH COURTESY OF TATE GALLERY, LONDON/ART RESOURCE, NY. © TATE, LONDON 2004.

*a Crucifixion*.... These figures ... were, according to Bacon ... visualizations of the Furies in the *Oresteia*...."[4]

What sense can we make of this strange creature? The references to Hitler and to the Furies tell us a little. In Greek mythology, the Furies were terrible spirits who punished the doers of unavenged crime, whereas Hitler, if nothing else, was a doer of crimes that were beyond avengement in this world. But these references do little more than reinforce what the artist shows us—the creature is monstrous. (As for who is getting punished here, and for what, we will see that before long.)

Consider that the beast has no ordinary limbs. Instead, we see those peculiar peg legs beneath its torso. They show us that this creature gets around—and got wherever it is in the world—by using artifacts. And on earth, we know there is only one species that makes its way—quite literally—with fabricated artifacts that fit its needs. The energetic brute can be viewed as Bacon's representation of what we commonly regard as one half of our human nature, the worst half. The artist has made a low-end picture of what we see as our grabbing, grubbing, self-serving *material* nature. That's the half that—when we seek to put down others or ourselves—we say is the *animal* half of our nature. True enough, lower animals live only in the present; they don't have good enough heads to recall the past or imagine the future. Like the wretched monster, active animals mainly seek to fill their faces.

At this point we must ask: What is so monstrous about eating? To understand Bacon's answer, we must pause to look at how we make monsters of ordinary people. We do two things, only the first of which amounts to much, and—voilà—the chore is accomplished. First, we degrade them. Then we get them to think of themselves as degraded.

For example, we know that it is the deliberate practice of pimps, when they acquire *potential* practitioners of the sex-for-money trade, to force sex on them. Just see that someone rapes 'em a few times, forces sexual entry into one orifice or another. That's the first task—it degrades them. The second task doesn't take much doing—the victims can usually be relied upon to lower themselves in their own eyes. No need now for the whip; they are willing to prostitute themselves. However we get forced to live degraded lives—fight the idea though we may—we think of ourselves as debased, and we soon find ways to actively act accordingly.

Bacon's allegory is that people living in the Western world through the middle half of the twentieth century were *changed* by that harrowing experience. Initially in wartime, we degrade the enemy, as they do us. But finally, beastly war with fellow humans, especially brutal war after brutish war, degrades all. The artist's powerful story shows that people who lived through those long years in the middle of the century could not persist that they belonged to a species of *enlightened* creatures. Lowered in their own minds they were—lowered to the grimy depths. In Bacon's untitled portrait of a passionately intense open-mouthed beast, those with time for reflection could see a fallen human nature. Twentieth-century hard times brought out the flesh — which was more than willing, it was eager.

To see the other half of Bacon's allegory — that the spirit was weak during the same hard times— we move on now to an example of the second of the two kinds of his pictures, *Study after a Portrait of Pope Innocent X by Velasquez*, 1953 (Figure 20–2). It shows us a religious leader, which is to say a person of spiritual authority. The artist portrays a blood-spattered pope sitting with his jaws ajar and his fists clenched. Unlike the rough beast, the spiritual leader seems contained if not imprisoned by a structure, in his case a tubular one. No doubt, the artifact that restrains this representative of the incorporeal spirit came from an assembly line, an industrial pipeline right next to the one that provided long peg legs for the meaty beast. In tough times, technology is our means to a variety of bad ends.

Sylvester notes that, in Bacon's similar *Study after Velasquez*, 1950, "the figure is at once monumental and evanescent.... The controlling force is the immensity of the mouth. It seems a silent mouth, open in a snarl, a realization of the desire ... 'to make the animal thing come through the human....' [Bacon's usage of] the long folds in the grey background curtain ... which he called 'shuttering,' was to formalize the folds ... into stripes that passed ... through a figure."[5]

Curtain folds that pass *through* the figure make Bacon's pope a suitably immaterial-seeming representation of the spiritual. But what about a mouth that snarls to make the animal come through the human? More likely (from a pope as representative of human piety) is a mouth trying to suppress that carnal prospect. In any case, the mouth is certainly not consuming anything. Why then is this pope's mouth open? When asked about that, Bacon said: "You could say that a scream is a horrific image.... If I had really thought about what causes somebody to scream [the painting might have been] more successful...."[6]

The portrayed pope is screaming, but Bacon is not explicit about why. So let's judge for ourselves. Would any proper spiritual leader in the early 1950s— aghast and alarmed when a fierce cold war was starting on the heels of the last hot war's end — not feel like screaming? Yet don't we expect much more than screaming, especially from someone who has a perhaps closer connection with what is believed to be God's almighty power than the rest of us? That may be precisely the reason the artist made so many portraits of a seated open-mouthed pope. They suggest that even someone like the bishop of Rome was powerless to stop what seemed like steps toward the destruction of our kind. Bacon paints a pope who is— at best — ineffective in the use of his spiritual authority. Scream though he may, the

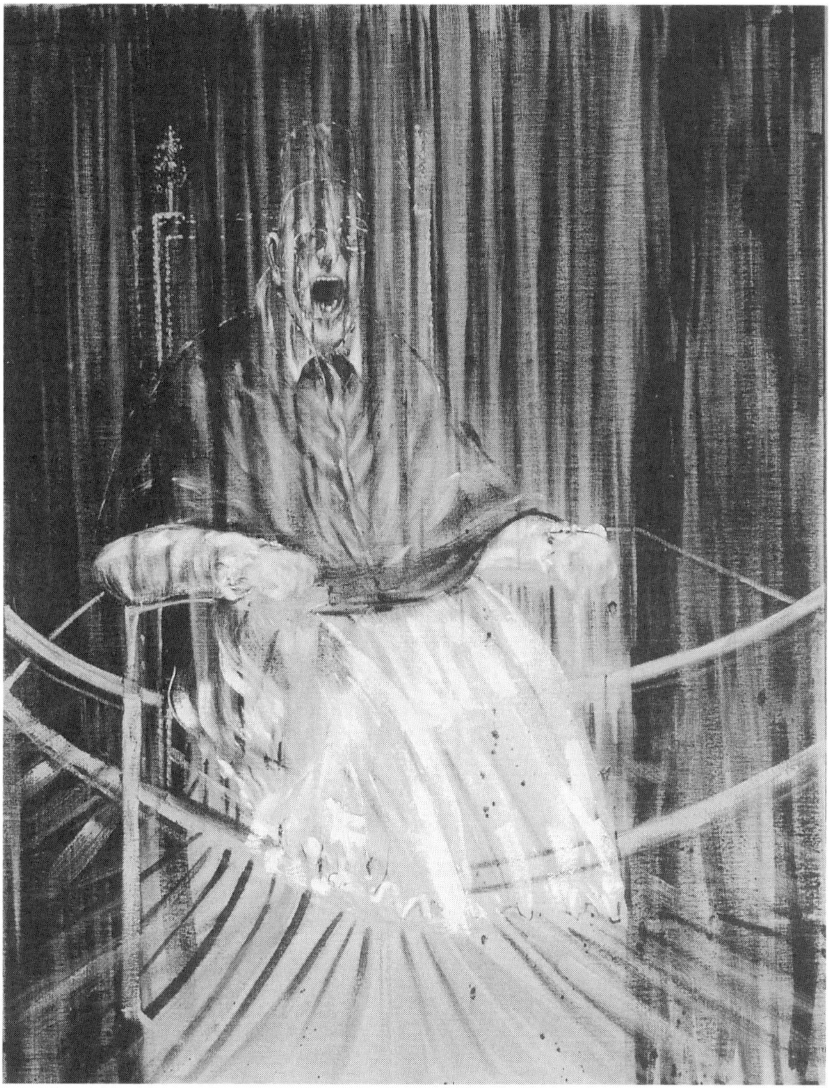

FIGURE 20.2. Francis Bacon (1909–1992). ***Study after a Portrait of Pope Innocent X (1574–1655) by Velasquez,*** 1953. XIR154923. (Oil on canvas) Private collection/Lauros/Giraudon/Bridgeman Art Library. COURTESY OF ©2004 THE ESTATE OF FRANCIS BACON/ARS, NEW YORK/DACS, LONDON.

man seated in the pope's chair in Bacon's *Study*— like others within or without the church — lacks all conviction that he can stem the flow of blood and the degradation of humankind.

Bacon's pope paintings express a despairing view of the other half — the spiritual half — of human nature. They show that the moral higher side within us all is, in a word, *impotent*. The spirit is simply no match for the flesh. This is another way the artist illustrates the same idea: Twentieth-century people — those who lived through the decades of brutal dictatorships, ruthless war, and inhumanly overwhelming weapons— had been debased.

Francis Bacon's beast and pope pictures are parables. They provide, each in turn, punishing and harsh portraits of one creature, the spirited human animal in hard times. The artist's paintings of the flesh part and the spirit part show that we were degraded in the twentieth century. He also tells us who defiled us. You know who it was. Especially in those mid-century decades, we met our match. Indeed, we met our enemies. They are us.

## *Chapter 21*

# Embodiment Without Bodies: Mark Rothko

It takes no more than a glance at Mark Rothko's *Light Red over Black*, 1957 (Figure 21–1), to be aware of his way of abandoning realistic *representation* in mid-twentieth-century art. The painting itself shows (using words that tell you next to nothing about it) a more or less rectangular patch of modulated color in the midst of not much else. Likewise, the small black-and-white illustration carries little of the picture's "presence." Without the color or the size of the artwork, it conveys only something of its shape.

The word "representation," however, means not only a portrait but also an "embodiment." So perhaps Rothko created an embodiment of some idea in his subjective mind. That is likely to be true. To be art, the picture must represent some clear or murky conception that was in its painter's head. But what about the heads of those picture viewers who don't see things like Rothko does? Whatever odds and ends it embodies for them, many may find *Light Red over Black* absorbing. So the very idea of what a work *represents* becomes problematic. Art certainly needs individual artists with ideas to create it, but after that it may take on an independent life of its own.

Rothko was one of the first modern artists to shun straightforward representation. That idea, however, was "in the air" for decades before Rothko and some other Abstract Expressionists made it their own. For example, you might look at Kazimir Malovich's decades-earlier *Black Square* (about 1920), but don't try to look at it here — it is not provided as an illustration. This writer's powers of description are good enough to convey its conceptual essence in words: Malovich's picture shows what a black square looks like.

As long as we are in this small mental realm where things may be a bit odd, try on this two-word Freudian interpretation as part of why art began to disdain and deny the objective realm of objects: science envy. After all, science had been going from one success to another in making large and small sense of the objective world. In fact, it had even provided the technology to do what once only art could do: make literal representations. Photography had become the new kid on the block of depiction. That was one of the reasons that many artists went more inward and non-literal in their depictions of people and objects.

FIGURE 21.1. Mark Rothko (1903–1970) ***Light Red over Black.*** 1957. Oil on canvas. 232.7 × 152.7 cm. Tate Gallery, London, Great Britain. ©1998 Kate Rothko Prizel & Christopher Rothko/Artists Rights Society (ARS), New York. PHOTOGRAPH COURTESY OF TATE GALLERY, LONDON/ART RESOURCE, NY.

Rothko himself saw no end of significant connections between his paintings and the state of the objective world. To comprehend those connections in Rothko's head, however, may have been beyond the capabilities of Freud himself. (Saying that may do Freud an injustice, as it goes along with our diminishing view of him over the last half century. Perhaps his father, Jakob, saw into the future when he said, "The boy will come to nothing."[1] To be sure, the father spoke those words only after the eight-year-old Sigmund relieved himself in his parent's bedroom.)

It was in 1950 that Rothko "developed the classic image of luminous, fuzzy rectangles of color that was to occupy his work on canvas for the rest of his life.... The development of his work was a whittling away of what he felt to be extraneous, an aspiration toward aesthetic essence in which pictorial content was reduced to volume, tone, and especially color. In this way, Rothko hoped to attain an art that would perform a kind of sacramental function, by embodying and clarifying the tragic condition of being."[2] Rothko sought a *sacramental* function with his art? Like, wow! Of course, what's going on in any other mind, especially if it's a creative one, can be hard to fathom.

For reasons of its own, the art community trumped even the artist with exalting, exulting prose. What generally greeted Rothko's work "was a stream of transcendentalist rhetoric.... A fair example: 'Unlike the doors of the dead which were meant to shut out the living from the place of absolute might, ... these paintings—open sarcophagi—moodily dare, and thus invite the spectator to enter their orifices. Indeed the whole series of these murals brings to mind an Orphic cycle ... the artist descending to Hades to find the Eurydice of his vision. The door to the tomb opens for the artist in search of his muse....' The higher criticism ... echoed the statements Rothko made about his own art, in all their ... ambition and ... cloudiness."[3]

Such high-flying words have some awfully Orphic orifice-entering to elevate them. Sigmund Freud, why are you lying down on the job when we need your help? Let us call in critic Christopher Knight to substitute a calm post–Freudian opinion. "The frame of language around Rothko saved his work from the kind of analysis that might have argued that Rothko, far from being Yahweh's official stenographer ... was a painter, a maker of visual fictions.... Rothko's dilemma was that he wanted to employ the vocabulary of Symbolism ... to render the patriarchal despair and elevation of the Old Testament.... [One of Rothko's formal statements in the 1940s said,] 'We assert that the subject matter is crucial and only that subject matter is valid which is tragic and timeless.... Without monsters and gods, art cannot enact our drama; art's most profound moments express this frustration.'"[4]

From a lowbrow viewpoint, it can be said that the artist's painting is more praiseworthy than his ambitious words. Plainly stated, Rothko's format, "which he hit on in 1949 and repeated with minor variations for the next twenty years, gave him an excellent matrix in which to experiment with color. In effect, it abolished practically everything but color ... Rothko's [creations] ... are vehicles for color sensation...."[5]

Apart from whatever his paintings represent, we should credit Rothko himself with a lot of guts for risking his head on the communal block when he made the *first* of these newfangled colored rectangles. After the first such creation was accepted by the art community, it would have been much easier for him to make more of them. The credit for those successor works belongs in part to that community—for confirming his initially risky self-esteem. Credit also goes to Rothko's own body chemistry—which, after the initial success, would have supported him with higher doses of hormones like testosterone and serotonin.[6] This is hardly unique to this artist's body; it happens with all humans, and with apes and monkeys as well.[7] Such self-dosage with internally made drugs might be the result of a Darwinian adaptive mutation. That's because it keeps any winner in a contest (among humans, apes or monkeys) from having to risk a sequence of such fights for a while, as the others recognize and kowtow to his gutsy behavior.

It was intended here to provide the viewer with a look at another one of Rothko's colorful displays of what may be the transcendent sublime: *Number 14*, 1960. The illustration's necessarily small size, not to speak of its lack of color, would have done no justice to this artist. So you must remain without intoxication as you hit the road to Willem de Kooning's *Woman* in the next chapter.

## Chapter 22

# The Subjective Stone Age: Willem de Kooning

In *Visual Arts in the Twentieth Century*, we read that one of Willem de Kooning's 1950 work *Excavation*, "buries figurative motifs in a jigsaw of anatomical bits and pieces...."[1] Such bits and pieces might be hard for many of us to see in what amounts to a dig into de Kooning's own subjective mentality. So here we will dig into some more accessible figural creations of his. We will seek the significance of the strange and powerful female figures in his grotesque "Women" canvases. Steel yourself before you turn to de Kooning's *Woman I*, 1950–52 (Figure 22–1). Then take a guarded look. You'll see that she is in equal parts female and fierce. In a later picture of her, *Woman II*, 1957, she is still undeniably female, but ferocity, as such, is not evident. That's hardly because she has become docile or gentle, but because she is more of a demon. Ogres needn't make a show of their fierceness, it is undeniable.

Let's turn now to a disarming early picture of her: *Woman*, 1948 (Figure 22–2). A glance at it will show you that, in her younger days, she was hardly fierce at all. Indeed she was not only young, she was a bit erotic in her appearance. A second glance shows that she is younger — she is not nearly as long in the tooth as in *Woman I*. What makes her more erotic is that the shape of her breasts is more visible. Note, however, that in front of this young woman (who has an oddly tilted head, which makes her seem somehow out-of-touch), there is a large spiked object that would keep most of us away from her.

Who is this Woman that undergoes these changes? Please be forewarned. The case will be made that she is something we don't like to think

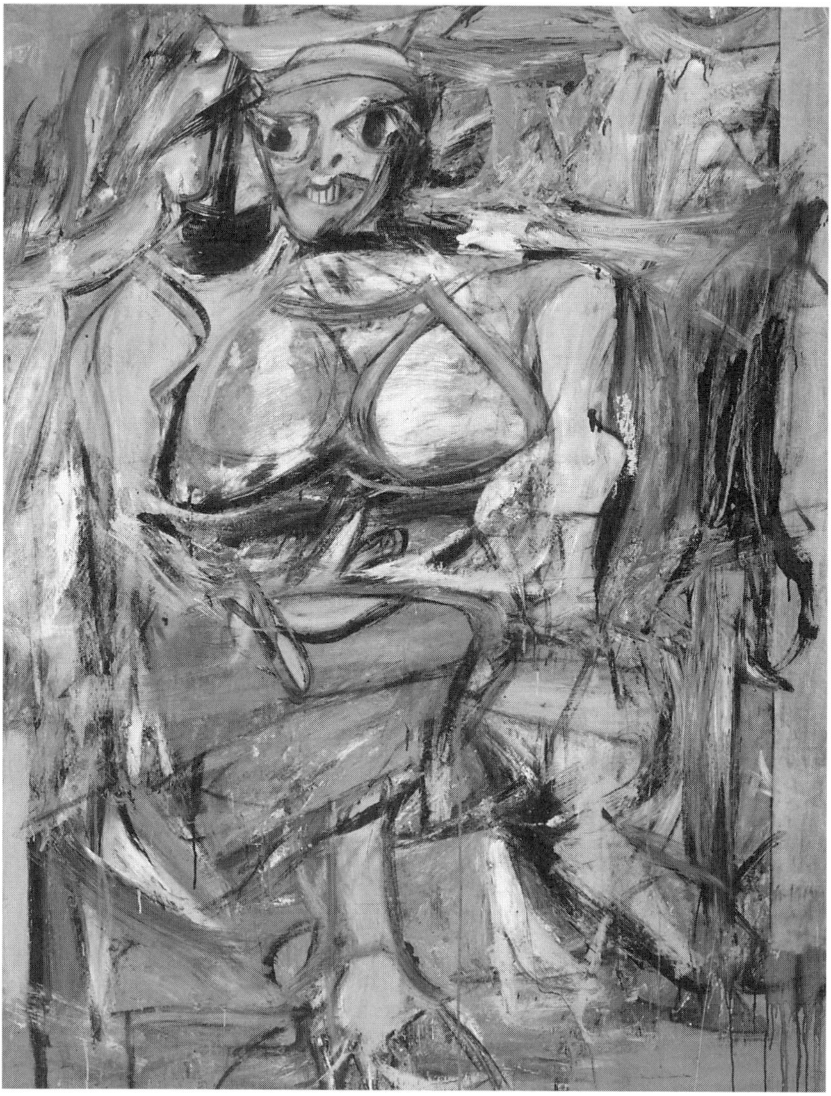

FIGURE 22.1. Willem de Kooning (1904–1997). ***Woman I***, 1950–52. Oil on canvas, 6' 3⅞" × 58". Purchase. (478.53). The Museum of Modern Art, New York, NY, U.S.A. ©2004 The Willem de Kooning Foundation/Artists Rights Society (ARS), New York. Digital Image ©THE MUSEUM OF MODERN ART/LICENSED BY SCALA/ART RESOURCE, NY.

exists, an "unnatural" view that sometimes comes to the minds of young children, especially boys. These pictures provide a child's-eye view of mothers, not as loving, but as ranging from attractive-but-detached to harsh, and then, monstrous.

Let's look first at some ideas about this Woman from others. Here are some words about her from critic Leo Steinberg: "In the spring of 1953, Willem de Kooning shocked the New York art world with six enormous paintings ... 'on the theme of Woman....' His *Woman* is ... no more distorted than ... a rainstorm is a distorted shower bath. She is a first emergence, unsteeped from a tangle of desire and fear, with some millennia of civilizing evolution still ahead of her.... Does that mean that ... she lacks 'human appeal?' Not in the least. In fact she is disastrously erotic in some remote paleolithic way."[2]

Writer Werner Haftmann also orients us about this Woman and her creator. He notes that being an Abstract Expressionist "does not preclude images of the outside world ... De Kooning ... [is one who] illustrates this.... In a long series of paintings De Kooning defined, as though to exorcize, this disquieting figure, erotic, semi-mythical — half-woman, half-demon — who appeared to him everywhere ... and forced her way into the abstract formal movement of his pictures."[3]

Disquieting (to say the least) de Kooning's Woman certainly is. Why she unsettles us will soon be made more clear.

Some of the painter's own words (in 1963) on the pictures provide help. "When I was painting those figures, I was thinking about Gertrude Stein, as if they were ladies of Gertrude Stein — as if one of them would say, 'How do you like me?' ... I look at them now and they seem vociferous and ferocious. I think it had to do with the idea of the idol, the oracle, and above all the hilariousness of it."[4] That de Kooning associated his Woman with writer Gertrude Stein provides us with a clue. Stein was outspoken and, somehow, fierce-looking. Certainly, her persona was "female," much more so than it was "feminine." However, Stein's image was not merely *female*, it was *matronly*. It is decidedly unnerving for a strange matronly woman to ask, "How do you like me?" Such a question has erotic connotations that might provoke an immediate nervous giggle as well as a later hilarious guffaw.

Psychoanalyst Carl Jung suggested that "non-objective art draws its content essentially from 'inside.' This is a complex area ... since psychoanalytic theories of art cannot be tested."[5] To say that something cannot be tested is to say that we are outside the realm of proper science. In this regard, Jung's remark is more candid that many of Freud's. Not that Freud is to be singled out for castigation for claiming that his views— such as the view that there was a universal romance between son and mother — were scientific. Asserting that odd views about women have the authority of science was also done by other early psychologists. For example, "the term *nymphomaniac*, to designate a woman with [what were thought to be] uncontrollable sexual desires, [was] a typical discovery.... Nymphomania

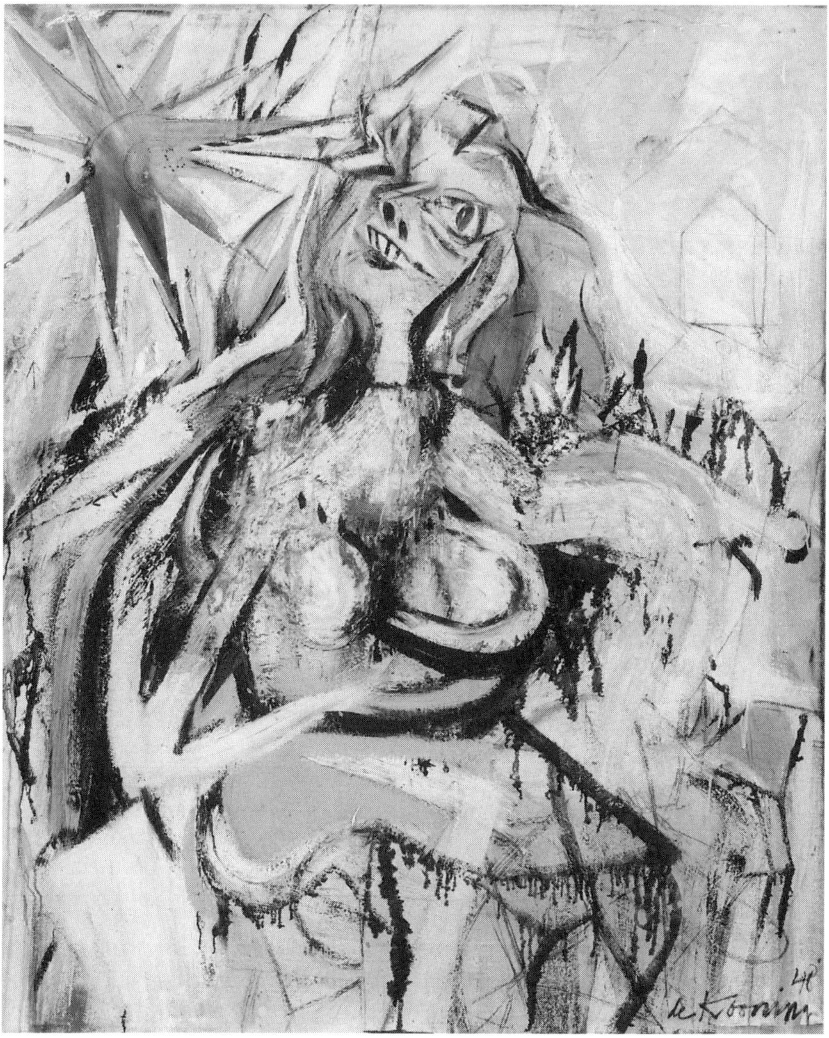

FIGURE 22.2. **Willem de Kooning. *Woman*, 1948. Hirshborn Museum and Sculpture Garden, Smithsonian Institution, Gift of the Joseph H. Hirshhorn Foundation, 1966. ©2004 The Willem de Kooning Foundation/Artists Rights Society (ARS), New York. PHOTOGRAPH BY LEE STALSWORTH.**

became a convenient and graphic term to describe a 'syndrome within the sphere of psychical degeneration,' as Kraft Ebing defined it in *Psychopathia Sexualis* ... which ... was characterized by irresistible exaltation and an insatiable appetite for sexual gratification."[6] Adding to such pseudo-science in those days was this notion: Masturbation was presumed by such authorities to be the common cause of this supposed ailment of women. Clearly, social science can be more easily tainted than can natural science by the odd beliefs of the times.

Despite the dire maladies that were thought to result from one-on-one sex and also from the do-it-yourself kind, chances are that — with and without partners—countless young nineteenth- and twentieth-century Jills and Jacks got up the hill and off, and did so with no harm (ignoring here sexually transmitted diseases coming from afflicted partners).

A real but less-than-Freudian connection between the de Kooning Women and their sexuality is something we will explore in a moment. First, some justification is needed for what may have seemed an off-the-wall opinion that his Woman is none other than a decidedly strange view of *Mother*. His Woman is certainly no modern emancipated feminist. One glance tells you that de Kooning's Woman had no need to be liberated; she was never subjugated. Neither is she a woman who ever had to resort to feminine "wiles," let alone a woman who had to run from a boss who was hoping to hit hard on her.

If Woman is not any of the above, what's left for her? What's left is that she is an unrealistic image in the sense that she may not exist in the public objective world (except perhaps for overblown views of a few national and corporate leaders). What remains for her is the subjective world, in particular the subjective world of the relatively unrealistic people called young children.

*Woman I* is a woman who is beyond defeat. Such a woman can exist only in the eyes of her young children. As for the later *Woman II*, she is not merely fierce, she is overpowering. She can be that only in the eyes of a would-be rebellious young child. To use de Kooning's term, it is *hilarious*— the impossible wish fulfilled — to think anyone can control her. But you can't tell that to the fantasizing toddler, who might like to get close to the earlier *Woman*, and might wish to find a way around the later ones.

We can agree with critic Steinberg that de Kooning's Woman is an emergence. It is likely, however, that she has emerged from the paleolithic past only in the strange sense that the development of the individual is sometimes thought to recap the life of the species. She comes from the *subjective* stone age — the secret minds of young children, for whom she was at once the whole world and mother.

Now we all know that our mothers are right up there with apple pie in the pantheon of frequently idolized ideals. We also know that mothers are merely human. Likewise, even adorable tykes are at times brats and pains in the butt. When they can't get their wants, they have far more in the way of egocentric temper tantrums than do the half-grown children that many of us retain in our heads.

In her earlier portrait, as *Woman*, she has an erotic quality that later gets dampened. It is this early picture that suggests she is seen through the eyes not of children in general, but of boys in particular. That's because there is *something*— no doubt less than Freud suggested, but something — to the idea that young boys sometimes look at women, including their mothers, as flushed with illumination from erotic floodlights.

To talk of a young son seeing a sex object in his mother is to remind us once more of the Freudian dangerous-family-romance story. For many of us now, Freud's Oedipal account is more a *tale* than it is something calling for our *belief*. To account for some of the peculiar thinking of males (and perhaps his own thinking in particular), Freud *created* the family romance by cutting a new pattern from the ancient cloth of Jocasta's tragic accidental marriage to her son.

Another reckoning of the roots of *man*kind's and *woman*kind's sexual thinking is a modern Darwinian one, which we currently think is more objective. The two sexes have disproportionate investment in sexual activity. The female — with the large investment of making a relatively large egg, and then the much larger investment of nurturing the baby before birth and after — is given *caution* by nature. The male — with low-investment seeds, which cost him little to scatter broadly — is, by nature, more *carefree*. Period, but not the end. These differences got built into our genetic material long before we got to be human. Then they got assembled into our human concept-making brains. Finally, they get *expressed* as subjective feelings and ideas. As females commonly seek sex less frequently, and as more opportunities than the ladies care for usually come their way, most do less thinking about getting sex than do most men. Endless such thinking — except for the times when they're busy making a living or when they're briefly satiated — is common among the gents. So, long before they grow into gents, and long before they realize such thinking might be unsuitable, the boys will be, well, males.

Art, once it is created, should stand on its own. So, although how this painter's art relates to his own background is of interest, that connection is secondary for his pictures' viewers. Nonetheless, it is at least interesting to learn that de Kooning "was raised by an extremely dominating mother.... He [was] out of touch with her for many years, only resuming contact when ... he could feel, at long last, that she no longer had any power over him."[7]

Here is a summary of this chapter's view of the de Kooning Women. In these incarnations that he gave her, Woman is none other than a young child's sometimes view of *Mother*, otherwise known as mommy, mama, mumsy, and a matriarch. She has emerged from the childhood minds that respect her, fear her, even hate her, and, yes, sometimes desire her. Those are the secret minds of *boys*, and of men (who still have boys within them).

## *Chapter 23*

# The Resemblance to Meaning: Lee Krasner

Take a look at one of Lee Krasner's early paintings, *Self-Portrait*, 1930 (Figure 23–1). It shows her painterly skills when she was still schooling herself, before she started to make Abstract Expressionist pictures, before she met and married her fellow painter husband (who just may have lacked the skills to make a recognizable self-portrait).

We enter a far different world when we turn now to a much later work of Krasner's, *Abstract #2*, 1946–48 (Figure 23–2). What is this picture about? Writer Ann Wagner suggests that this and related pictures of Krasner's "use their means to real rhetorical purpose: they aim to negate any notion of 'language' in painting.... Its measure stays even grained; its marks, though separable, are equivalents of each other; they seem, if not exactly anxious *not* to mean, then at least reluctant to do so.... Meaning remains at the level of a resemblance approached, then skirted once and for all."[1]

That the picture negates—rather that mimics—any notion of simple language is no surprise. That is something we expect from any picture worth its paint. After all, if ordinary communication with words could do the job by itself, why mess around with pigment? That's so, even though the art movement called Conceptual art sometimes made use of ordinary words. It was "essentially an art of mental patterns ... its basic material was ideas...."[2]

However, for meaning in the painting to be skirted and avoided is something else, something that should make a viewer uneasy. Certainly a painting need not tell a story (such as the fun story we saw with Ingres' history painting *Oedipus explains the riddle of the Sphinx*). It could be, for example, a portrait. Krasner's early *Self-Portrait* neither mimics language nor tells a story. Nonetheless, that picture is meaningful, or at least — to use a couple of broader terms— it has import and significance. If Krasner had wanted to make a meaningless self-portrait, she could have either made a very bad one or cut a good one into random pieces and pasted them onto canvas. *Abstract #2* is neither of those.

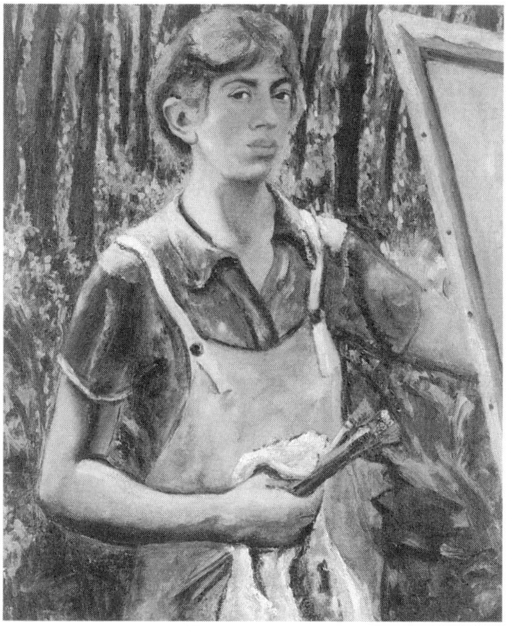

FIGURE 23.1. Lee Krasner. ***Self-Portrait***, **1930. Oil on linen, 30⅛" × 25⅛", 76.5 × 63.8 cm. ©2004 The Pollock-Krasner Foundation/Artists Rights Society (ARS), New York. COURTESY OF ROBERT MILLER GALLERY, NEW YORK.**

Yet there's more to art than story-telling and portraits (and landscapes and still-lifes). Wagner also describes "nonfigurative art [such as Abstract Expressionism] ... as a prime means for self-inscription.... [Such self-inscription is kind of a portrait, not of the artist's physical appearance, but of her interior self.] For such painting to occur, the particular conventions of depiction must be laced with the means to convey the directness and immediacy and idiosyncrasy that have come to be associated with the self.... The whole rhetoric of spontaneity ... [can] be a vehicle for the self...."[3]

We may be getting somewhere. Perhaps *Abstract #2* is a way Krasner chose — not only to skirt straightforward meaning, but to show her unique subjective self. Writer Edward Albee sees such self-expression in some still-later pictures of Krasner's, such as *Icarus*, 1964 (Figure 23–3). He writes: "What strikes me most about these paintings ... is their exuberance.... These paintings dance and prance and sweep and swerve and shout in their enthusiasm. All is exuberance — the great arcs of paint, the astonishing juxtapositions of color.... What it is, most precisely, is the joyous outpouring of a major artist who is sure of her gift and is happy to share it."[4]

No doubt, looking at the picture will make many agree with Albee. Others may not. This writer must confess— he is one of those who just

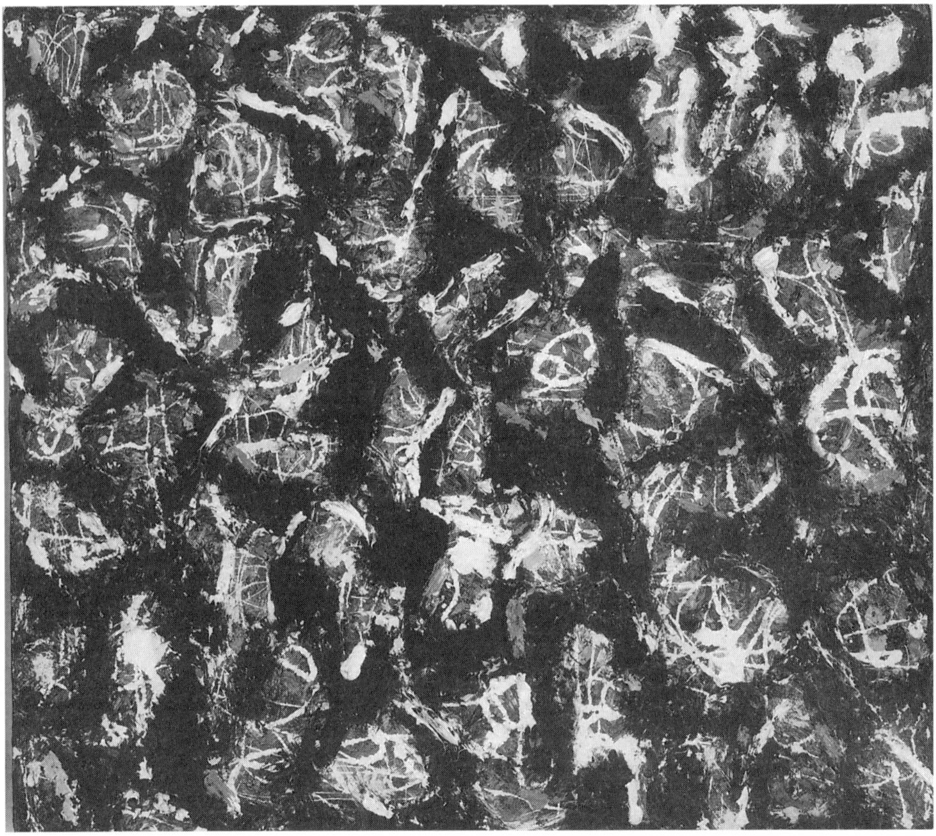

**FIGURE 23.2. Lee Krasner. *Abstract #2*, 1946–48. Oleo sobre lienzo; 52 × 59 cm. Institut Valencia d' Art Modern, Generalitat Valenciana. ©2004 The Pollock-Krasner Foundation/Artists Rights Society (ARS), New York. PHOTOGRAPH BY JUAN GARCIA ROSELL.**

don't perceive the exuberance when he sees an image of *Icarus*. What he learns is something altogether else. He learns once more that art is like religion. This is not a reference to the spiritual uplift claimed by many partisans for each. Rather it is this—both art and religion have a large subjective component. In fact, it may have been art's long-ago task to *show* us how the subjectively experienced supernatural world "will flame out" from the natural world, "like shining from shook foil." (The quoted words are borrowed from Gerard Manley Hopkins's poem "God's Grandeur.")

In truth, the appeal and understanding of specific religions, and of particular artworks, is not as widespread as their devotees would like to believe. So this writer's own subjective nature may be one of those not

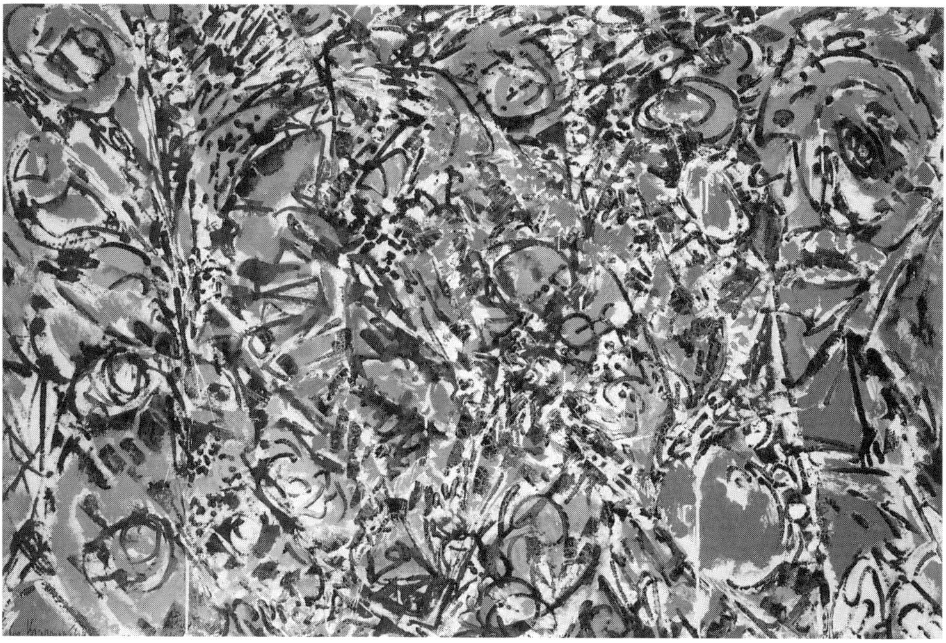

FIGURE 23.3. Lee Krasner, *Icarus*, 1964. Oil on canvas, 46 × 69 in., 116.8 × 175.3 cm. ©2004 The Pollock-Krasner Foundation/Artists Rights Society (ARS), New York. COURTESY OF ROBERT MILLER GALLERY, NEW YORK.

shaped to swear allegiance to certain deities, or — getting to the point — to see Krasner's joyous spirit embodied in *Icarus*.

As long as he is in the confessional, allow this writer to admit some grosser limitations. Instead of accepting that his own understanding of Krasner's subjective paintings is limited, he sometimes places the flaw in her and wishes she had continued with her earlier mode of painting. In fact, he wishes that not just her variety of it, but all of Abstract Expressionism would just go away. Thus he is tempted, by some of the eerie ideas of modern physics, to indulge in a grandiose fantasy where he can go back in time and change things. Not to fix all the world's problems. No way, he just turns off Abstract Expressionism.

Here's how this writer does it. He takes advantage of some of the very latest we have learned from science, as described in *The Non-Local Universe: The New Physics and Matters of the Mind*.[5] That revolutionary news is that 1997 experiments with photons confirm what was suspected — that our universe, the whole thing, is what's called *non-local*! What that means is that every last particle in the universe has been "entangled" with other particles since everything started. Thus, modern physics says, we must "abandon the mechanistic or classical concept of causality...."[6] Abandon causality, causality that puts the *forward* direction on time with an arrowhead? Now, that is revolutionary! Indeed, that can be taken to mean that tiny quantum wave/particles behave like there sometimes is *backward-action* from the present to the past! The new quantum physics nonetheless insists that *information*, as such, cannot be sent back in time. But something in the way of *influence* can somehow go from here-and-now back to there-and-then.

Now here's this writer's fantasy. What he does is send not some information (he knows better than to try that), but something else — some rhetorical influence — back to Krasner and her fellow soon-to-be Abstract Expressionists in early 1950s New York. It might not be easy to achieve backward-causation, so he decides to send only three influential words. But, in truth, he has no idea how to do that objectively. So, what the heck, he tries to send them subjectively. How? Lacking any tool for that, such as a wishbone, he simply closes his eyes and wishes back to Krasner his hopefully powerful words: "*Don't go there!*"

Now that he has exhibited his own megalomania, allow the writer to point out that some modern physicists may have what could be called their own mania. They have given us the "many-world interpretation" of the nature of everything. For anyone who is a "glutton [of ideas about the nature of reality] there is nothing to touch [the] so-called 'many-worlds' idea ... of an infinity of parallel Universes...."[7]

With no supporting evidence, nor with a good idea of how to get such certification, we are told that our non-local universe may be just one of an uncountable number of such universes. Grab hold of something to support you: There's more than one universe, there is inestimably more! Much more astonishing: Such new entire universes are continually coming into being! Zillions of them. According to this astounding notion, "the room in which you now sit is splitting into virtually identical rooms with virtually identical observers billions of times per second."[8] No doubt there is some value to this belief, perhaps something like this: It allows scientists to avoid what is distinctively not comfortable, the idea of something that is *one-of-a-kind*, and replace it with what, at least in principle, can be handled, a *plurality*.

Well, wishing doesn't seem to have worked with getting Krasner not to go where she went in our world. So, just in case weird world-splitting is the way things really are, this writer decides to take advantage of that wild idea, like this: His carefully crafted message ("Don't go there!") did actually result in the sending of *one* lousy little photon back in time. There, it split off another entire universe. That allows the writer to think his mental efforts to turn off Abstract Expressionism are paying off right now. Not here, he knows that! It is working in a gul-danged parallel world.

You may not grant him absolution for his above admissions, but this writer does feel a bit better. More important than his present feelings is this: Wanting to feel better back when Krasner worked must have been a large part of the reason for art's inward turn in the frightening mid-twentieth-century world. Back then, we were losing our religion — which went by the name of belief in the beneficial progress of science and technology. Back then when an arsenal of a few atomic bombs was being replaced by huge opposing stockpiles of far more powerful hydrogen bombs, comfort could be gained by turning to the inner life. And art pictured what such turns are like. Abstract Expressionism may not provide comfort for all who looked at its pictures. But that's hardly a fair rap against it. After all, even old or new religions provide comfort only for some seekers of consolation.

To get back to Krasner, one curious thing was her reaction to the de Kooning *Women* pictures (we saw some in the last chapter). Two decades after she first saw them, she said, "With regard to his series on *Women*, I reject them one hundred per cent; I find them offensive in every possible sense; they offend every aspect of me as a woman, as a female.... It is the hatred and hostility toward the female...."[9] Perhaps she might have moderated that view if she had considered the last chapter's interpretation of the pictures: They represent the only occasional viewpoints of very young boys toward their mothers.

Lee Krasner lived and painted in a far different world than did fellow artist Berthe Morisot some seven decades earlier. Yet there is one way in which Krasner continued to live in Morisot's world. Albee tells us about the reaction to Krasner as a *female* artist: "Even a few years ago ... there was so much extra-art garbage surrounding [Krasner's] work.... These piles of garbage were named 'female artist' and 'wife of artist....' The notion that women are less aesthetically profound and innovative than men ... doubtless spreads back to our beginnings as upright animals.... The theory [that] men provide, women stay at home, cook and rear ... began to break down seriously only in this century...."[10] Morisot had to put up with critics making what they could of the fact that her husband's brother (Edouard Manet) was also an artist. The same goes for Krasner, whose husband (Jackson Pollock) was also an artist.

Here's one little story that shows how she put up with an interview, and with her hubby. She was asked about her husband's reaction to her work. "Very sympathetic.... He did not criticize [or offer suggestions]. He simply did [the following on one occasion:] ... I was out one afternoon, and I come in, and I found that the painting I had kept on the easel, that I was working, [had fresh paint on it] ... and I said, 'That's not my painting,' and then the second reaction was, he had worked on it.... In a total rage I slashed the canvas ... and I guess I didn't speak to him for some two months."[11] Let's not try to imagine how her husband (who was given to boozing and then being a bit of a brute) would have reacted had she—deliberately or accidentally—dripped or dropped some paint on one of his works.

## *Chapter 24*

# The Meaning of Meaningless: Jackson Pollock

Although the drip paintings of Jackson Pollock were and are controversial, no one disputes one aspect of his earlier work. He was neither much of a craftsman nor good at the expression of feelings in paint. "In the 1930s, Pollock's closest supporters charitably indulged his crippling deficit of facility. (A kind advisor ... looking at his early work ... would have suggested he take up tennis or plumbing instead.) Around 1939–40, however, that wincing tolerance began to give way to genuine interest."[1] Pollock's shortcomings are not brought up to put him down, but because his lack of such skill just might be connected with his main product, the drip paintings. After all, none but Athena sprang entire and perfect from the brow of Zeus. Most of the rest of us got where we are as a result of our flaws as well as our finest skills.

After a while Pollock stopped trying to show figurative content. He went another way. Like the decades-ago football player who became famous when he picked up a midfield-fumbled ball and ran for a touchdown — ran the other way, thus scoring for the benefit of the other team — Pollock became a big-time artist by running the "wrong" way. Before exploring which way was the wrong way, let's look at some of Pollock's paintings.

In the 1930s and 1940s, he made pictures that showed real objects. During the middle of the 1940s, his country fought another world war, a war that ended with the use of new weapons of previously inconceivable power. His picture *War*, 1947 (Figure 24–1), presumably reflects on that big war. It shows a human and perhaps a cow falling onto a heap of weapons and bones. Whatever its shortcomings as a moving work of art,

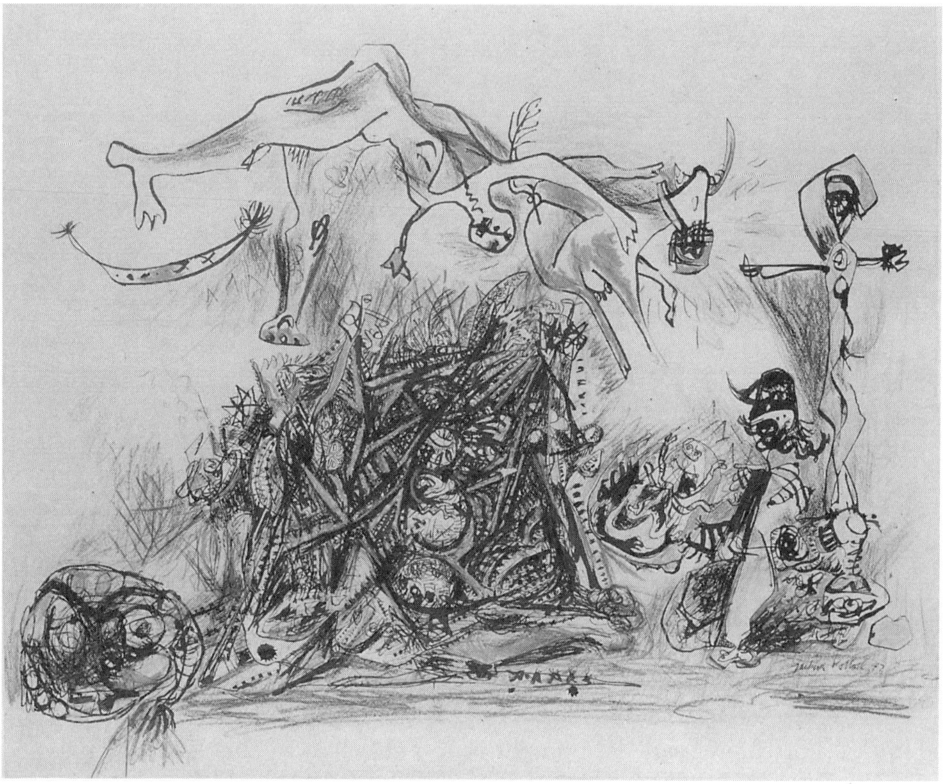

**FIGURE 24.1. Jackson Pollock (1912–1956). *War*, 1947. Pen and ink, colored pencils on paper. 20⅝ × 26 in. All rights reserved, The Metropolitan Museum of Art. Gift of Lee Krasner Pollock, in memory of Jackson Pollock, 1982. (1982.147.25). ©2004 The Pollock-Krasner Foundation/Artists Rights Society (ARS), New York.**

they are surely magnified when we compare it with Picasso's compelling *Guernica* (Figure 16–4).

Unfair to put Pollock up against Picasso, you think? It is done because Pollock did it himself. That cheeky self-esteem may have been another factor that contributed to his switch to dripping paint. "Despite the insecurities that otherwise plagued him, Pollock ... formed an arrogant cocksureness about his stature as a painter. He thought he was the best, out to 'kill Picasso.' [He was] incredibly sure that he was a great painter.... He thought he was the greatest painter ever."[2]

A way of deliberately but accidentally dripping paint in splotches and threads became the means for Pollock's claim to fame. To see one of the results, look at his *Autumn Rhythm*, 1950 (Figure 24–2). What you won't

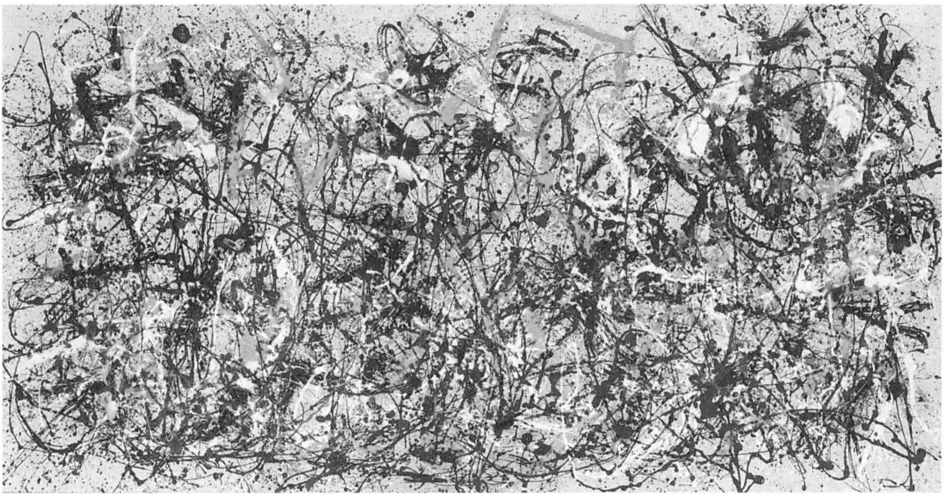

FIGURE 24.2. **Jackson Pollock (1912–1956). *Autumn Rhythm*, 1950. Enamel on canvas. 105 × 207 in. (266.7 × 525.8 cm.)** 

see are *objects*. If it is not a new way of painting worldly objects that is significant about Pollock's mature work, then what does propel it? In part, the answer is that he abandoned the outside objective world so that he could go inside to somehow portray something of his own subjective life. Considering his over-the-top self-esteem as well as his painful self-doubts, painting from his swirling conflicted inner life does make some sense. We should also consider that Pollock had been influenced by the Surrealist movement, which sought to portray irrational aspects of subjective life. The Surrealist writer André Breton gave technical instructions on how to work. "Write quickly, without a preconceived subject, fast enough not to remember and not to be tempted to read over what you have written."[3] This suggests that clear meaning was not what Breton sought. The same can be said of Pollock.

Here is how the author of *Painting in the Twentieth Century* puts it: "Pollock was ... interested in ... the general idea of spontaneous projection of inner states.... [He] looked upon the picture surface as a field of tensions between contending pictorial forces.... In 1947 Pollock set these automatic energies wholly free and arrived at an extraordinarily spontaneous and direct script.... The dripping paint follows the sweeping body movements of the painter. The picture becomes a kind of psychomotor choreography."[4]

Whatever Pollock's own motivation for making it, what does a picture like *Autumn Rhythm* do for those who have insides not shaped to perceive what Pollock's could? For them, here's a suggestion: The picture is like a Rorschach inkblot. A Rorschach inkblot is the basis for a test that allows individuals to project their own meanings into a pattern that has no meaning of its own. So perhaps each of us can find some private meaning in such works. Yet that does not explain the broad impact of his pictures on the art world.

There may be indirect help from some comments (offered decades ago) by critic Leo Steinberg. "More than any other living artist's, Pollock's work has become a shibboleth.... Arrayed against the sheer validity of Pollock's work are the views of most art historians, of most literary men, and of many brilliant painters not in the *avant-garde*. Pollock's champions are a few critics and museum men, abstract painters who recognize in him a superior power, and, above all, those who know the man himself."[5]

That sounds like Pollock not only had enormous belief in himself, but was able somehow to put that presence into his paintings. It also sounds like the few who followed the leader began to acquire enough of his charisma to overcome the doubts of many of those who didn't believe. After all, between Jesus and Christianity stood his disciples and their followers. Even with Jesus, however, it was not merely the special quality of his leadership that made the good news spread. It was the ever-so-meaningful message, the idea that he died so that those followers and others could be saved and might live forever. Not that we should expect Pollock's message to be at all comparable with that of Jesus, but let's ask: What, pray tell, was his powerful theme?

About meaning, writer Paul Wood provides indirect help: "All works of art mean something. They are the product of purposive human activity."[6] Wood also notes that "Many accounts [of Pollock's works] have been offered. These include: [1] formal configurations appealing to eyesight alone; [2] traces of anxiety at the onset of the nuclear age; [3] a sense of both achieved aesthetic unity and its irrevocable loss...; [4] traces of hedonistic physical joy; [5] metaphysical meditations on being; [6] an attempt to cope with urban life; [7] an elevation of the process of making over the finished product."[7] The variety itself of these meanings suggests that Pollock's drip paintings are indeed like Rorschach inkblots—different strokes for different folks. Yet multiple private meanings, however suitable they are for Rorschach inkblots, are not enough to account for these works that had so much effect on the art community. There is reason to think they also had a big overall meaning.

Intended by him or not, Pollock had an overall message that was powerful enough to convert much of the art community, although it was scarcely the kind that would move the masses. Pollock's message was less good news than it was bad news. His strange message was hardly that the meaning of life was to be found in the Father, the Son, and the Holy Spirit. There is a simple but powerful message that presents itself to most of us who view *Autumn Rhythm*. It is the very same message as the objective meaning in a Rorschach inkblot. Pollock's message was not that the meaning of the outer world was lost or difficult to find; it was that meaning of that kind didn't even exist. Thus his drip paintings boldly and blatantly proclaim: "*There is no meaning out there!*" That was the way wrong-way Pollock went when he couldn't do well at making meaningful pictures of the outside world.

Critic Kirk Varnedoe has some confirming words. "Pollock had spent over seven years wrestling with Picasso and a few others, in an effort to charge his paintings with meaningful content and potent symbols. Those struggles would now be left in abeyance.... Changing his self-awareness from a search for buried icons or totems to a reliance on more pragmatic instincts about how it felt best to work, Pollock would unblock the way to a fundamentally personal, original art."[8]

Getting unblocked must have been a relief for Pollock. The resultant drippings and droppings of paint were enough to fill large canvases like the nearly nine-foot by eighteen-foot *Autumn Rhythm*. Why were they made so large? To provide a way of *insisting* on their significance, despite their status as objects that lack meaning. Varnedoe also writes: "Pollock in 1947 ruptured the existing definitions of how art could be made, and offered a new model of how one could be an artist. Spontaneity and chance ... were now brought forward into daylight as tools of concrete engagement with materials and the act of making...."[9] To further pound at the point about Pollock, haphazard activities are just not meaningful.

All that may have been fine for Pollock, and it showed more than his self-esteem; it showed his courage in entering such a bramble-filled void. But why did his doing what he did so excite the art community? After all, no man is an island. Nor is any painter acknowledged as an artist until he's accepted by the main community of those who decide what is art. Let us examine what such stunning success—with the meaningless results of fierce chanced drippings—says about the art community and about our Western society.

The insistence that it has major meaning sends millions of Hindus into the waste-laden Ganges River once in a lifetime. It regularly sends far more than mere millions into synagogue, mosque and church. Among the latter, there were in Pollock's mid-twentieth-century America, as there are now, millions who cannot accept the evolutionary science that Darwin gave us. What they just can't abide is the thought that human life is the consequence of nothing more than natural selection from countless *meaningless* accidental mutations. The same insistence on meaning now makes many thinking people assert that we and other life on earth are the work product of an Intelligent Designer. Others suggest that the whole shebang of our present universe is *not* just the outgrowth of a meaningless big bang. For them, the Anthropic Principle—which says that everything was somehow *tailor-made* to become a fit place for us, but mentions no tailor—invests the whole thing with meaning.

If people of every kind fight hard *for* meaning in their world, why would a painter like Pollock win the support of the art community? After all, his works do more than hint; they proclaim that there is no meaning outside. The unpleasant but somehow exciting notion of a meaningless world comes to us—indirectly, to be sure—from science. Physical science, the source of most of the change in the world, doesn't blurt out, "There is no meaning in the world!" But it continues to make progress without satisfying the heartfelt human desire for a world with overall meaning. So the suspicion arises that meaning in the world may be neither important nor necessary, and it may in fact be absent. Also, art serves here as a glove that distances us from direct contact with the really hot potato—no meaning in our *own* lives.

To put this suggestion about Pollock's work in more dramatic terms, not only are his drip paintings lacking in external meaning, that lack *is* the meaning of such work. That is the truth about Jack the Dripper, but it is not the whole truth. Looking at that truth also provided the art world with a way of not looking at another truth. It was savory to dine on a hot potato while still having it as an art object to behold. Denizens of the art district were able to focus on Jackson Pollock's succulent drippings. In so doing, they turned their faces and minds away from a truth that was too hot to handle. They lived in a world whose central feature had become widespread death from forceful despots and powerful weapons. The art world could say, "Just don't look there, look here. Look how exciting this is!" That was the second main reason that the art community bought into Pollock.

Some people have suggested that Pollock's repetitious making of meaningless drip paintings implied that his tank was empty by the time he died driving drunk in an auto collision in 1956. That may not be the case. Among his last works were some that were other than dripped swirls of paint, such as *Easter and the Totem*, 1953. The author of *Jackson Pollock* sees signs in it of new life.[10] We would see a flower and a person, springtime signs of the rebirth of the old life of art, signs of the resurrection of meaning.

# PART VI

# Objects in the Mirror Are Seen as Shallow: Late Mid-Century

### *Chronology of Technological and Scientific Developments*

1950: Hydrogen bomb development starts
1950: Tranquilizer Miltown comes into wide use
1952: First hydrogen bomb detonated
1953: Lung cancer attributed to cigarette smoking
1954: Salk antipolio serum first used on children
1956: First transatlantic cable phone service
1957: *Sputnik*, first earth satellite, launched
1960: Laser device developed
1962: Thalidomide drug is found to cause severe birth defects
1963: First use of artificial heart during surgery
1967: Synthetic version of DNA produced
1969: Man first steps on the moon

## *Chapter 25*

# How Real Can It Get?

In the last few chapters, we looked at Abstract Expressionism as an inward-turning response to the frightening aspects of the endless changes that science and technology brought. After that inside move, the art world swiveled back to the outside — with Pop and related art. This shift is sometimes described as a change from *modernism* to *postmodernism*. It will be more useful to regard Pop and associated art as showing a different kind of reaction to the relentless progress of not only science and technology but of coarser forms of objective knowledge. Much of Pop art portrays the countless artifacts that had become so plentiful that we humans seemed to live embedded in them. The marketing of such stuff had become a large part of modern objective knowledge. In reaction, disdain for them and dread of the materialism they presumably represented became a big factor in the views of various cultural elites, including the elect called the art community. To that last elite, what was painful in particular was the expanded merchandising of art — it made art workers feel like kept sex objects. Nonetheless, art continued to have, quite apart from market value, artistic value.

A radical view of the connection between art and science is recalled for us by writer Brian Silver with "the words of William Blake: 'Art is the Tree of Life, Science is the Tree of Death' ... no ambiguity, no mincing of words."[1] So we should ask, did art turn from the inner life shown in Abstract Expressionism, and turn to showing science as death's representative? Although art certainly turned away from strong feelings, it pictured nothing like scientific versions of grisly dances of grim reapers harvesting haystacks of us. Nor did it turn to direct portrayal of how we feel about science itself.

With Pop art, the trick art turned was more complex. Art sought to portray human life as if it were little more than the activity of material objects about which science collects data and writes equations. Pop art painted material objects, especially those that humans make. It also pictured what seems decidedly odd—humans as mere material objects. It was odd, but it was exactly the way that some scientists began to depict human life. In *The Taboo of Subjectivity*, we read: "Since the Scientific Revolution, subjectively experienced mental events have gradually lost their status as real entities.'"[2]

Isn't that wild? How can people say such things about what seems more real than anything—our feelings, sensations and thoughts? Hey, it's not hard. They expand their beliefs about the importance and value of science until they fill the entire human world and become *scientism*. "Although signs of scientism can be found in writings as early as the seventeenth century, they have become far more prevalent since the nineteenth century with the rise of scientific positivism.... Its three central assertions are that 1) science is our only source of genuine knowledge about the world, 2) science is the only way to understand humanity's place in the world, and 3) science provides the only credible view of the world as a whole."[3] In the twentieth century, scientism took the form of *scientific materialism*, an "ideology that suppressed modes of inquiry into mental phenomena that do not conform to its principle.... [That] dogma would rather deny the existence of introspection, or at least marginalize its significance, than acknowledge that ... we still have no scientific means of exploring consciousness directly."[4]

Scientist Richard Lewontin cautions us that "biologists know what a brain is, but they are as confused as ever about mind.... Their sympathies are with the tradition ... that all talk of 'consciousness' is a metaphysical delusion on which we should stop wasting time."[5] Some put it more directly. Philosopher Paul Churchland says, "mental states should be regarded as nonexistent or at most as 'false and misleading.' [Philosopher] Daniel Dennett asserts: 'there simply are no qualia [roughly, immediate direct experiences] at all.'"[6] Writer Tom Wolfe describes an expansion on the same theme: "Neuroscientists involved in three-dimensional electroencephalography will tell you that there is not even any one place in the brain where consciousness or self-consciousness is located. This is merely an illusion...."[7]

This is not meant to suggest that assertions of that kind—that deny independence to subjectivity—are just plain *wrong*. Rather, what is intended is that simply asserting such challenging contentions—without the supporting evidence that is not provided—may be closer to scientism than to science.

Please note that there are many others who stretch out the authority and language of rational thought. For example, the proponents of Intelligent Design currently seek to sap the strength of the idea of Darwinian evolution. They believe there are complex life structures whose mysteries cannot be untangled within nature and natural selection. Their explanation — that such presumed unsolved mysteries imply the existence of non-human Intelligent Design — is peculiar. If their argument was accepted, it would not relieve us of mystery. Rather, it would replace down-to-earth mystery with the far greater mystery of an unearthly Intelligent Designer. A reasonable comment on that would be, "Why bother?" If an intelligent designer God exists, He may—for His own unfathomable reasons—have made a world without things that are difficult for us to explain, or made one with such mysteries. The plainest explanation for our inability to explain some things is that our abilities are just plain limited.

Be all that as it may, why would art and artists seem to be concerned with scientific materialism? Because "modern painting cannot be considered apart from the sum total of ideas developed in the 20th century.... It is a reflection of transformations in a deeper layer of humanity ... in man's whole attitude towards the Self and the World, Heaven and Earth.... Such transformations range over ... mental and spiritual life.... Over long stretches of time, [such changes] operate slowly ... then suddenly acquire explosive strength, and in a revolutionary upheaval take on an entirely new structure.... The beginning of [the twentieth] century marks such a radical break ... modern painting is one of its expressions."[8]

Around the middle of the century there were two more radical breaks for painting. Between them, the break to the inside world of feeling and experience and then the break to the outside world of objects and information would further widen the division in art. We saw pictures from the first of these in the preceding chapters. The chapters coming will show some from the second. We'll look first at the works of Jasper Johns and Robert Rauschenberg. Then two genuine Popsters, Roy Lichtenstein and Andy Warhol, will show us their variations on the theme. The works of these artists might really thrill you, or they could make you totally tranquil, like a machine. Perhaps thinking about the meaning of their art will excite you.

## *Chapter 26*

# An Art of Artifacts: Jasper Johns

The art of Jasper Johns inquires about our life in the midst of countless human-made material objects. Much of it focuses on flat kinds of such artifacts. Some of it confronts us with the idea that we humans may ourselves be mere material objects living lives that have no depth.

Flat human-made things—targets, flags, maps, digits, or alphabet letters—are what fill many of the works of Johns. A glance at *Flag*, 1955, or at *Figure 1*, 1955, would inform you that you are in the presence of a lushly painted version of the American flag or of our first numeral. Pertinent to the significance of such works is the remark of another artist (David Salle) that "everything in this world is simultaneously itself and a representation of the idea of itself."[1] However, things such as the targets, numbers and other flat objects (that Johns paints versions of) all have an additional quality: Their significance lies *little* in them as objects but *mainly* in them as representations of ideas. For example, consider the written numeral "1," and then the idea behind it. Clearly, just how the numeral is written matters little relative to the idea about its meaning. That suggests that Johns is entranced more by ideas about things than he is by the objects themselves. Despite that, the artist in him exhibits those interests with skillful, sensuous, two-dimensional representations.

Within some of Johns's artworks there are not only flat painted objects but also three-dimensional objects. For example, *Target with Four Faces*, 1955 (Figure 26–1), shows not only a round ringed target made with paint, but above it are plaster casts of the most significant part of a human body.

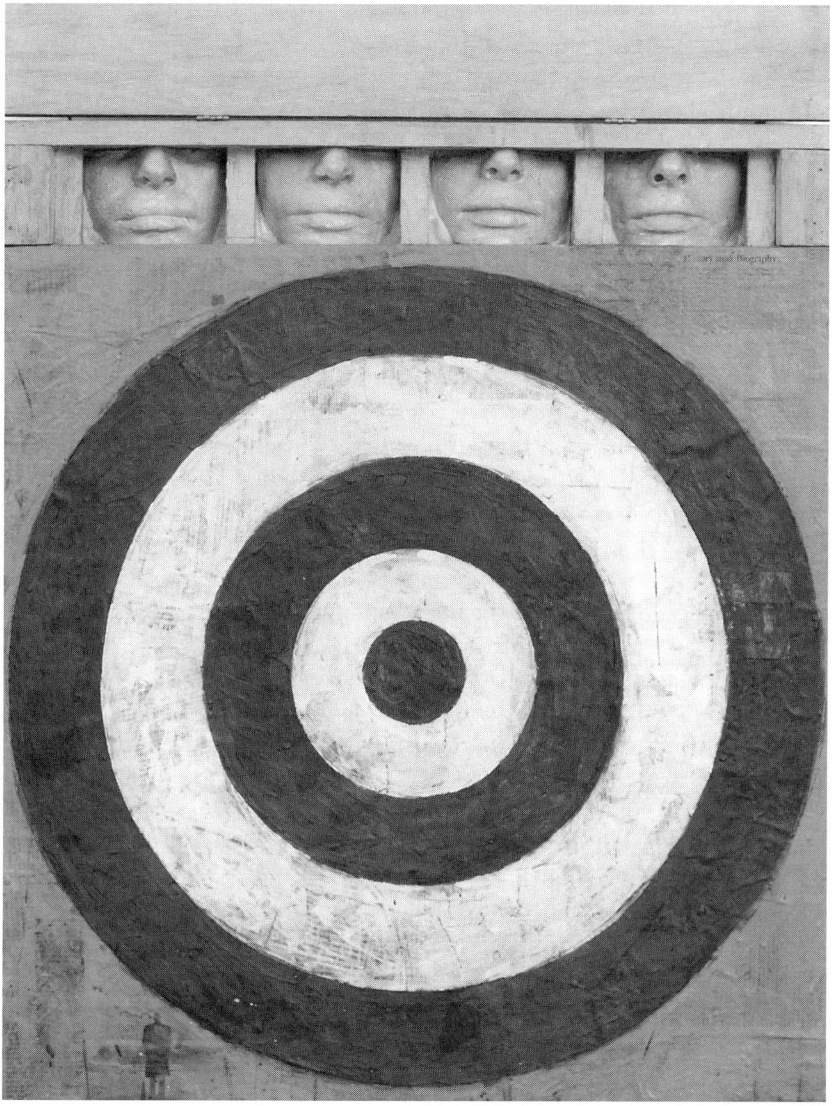

FIGURE 26.1. Jasper Johns (b. 1930). ***Target with Four Faces***, 1955. Assemblage: encaustic on newspaper and collage on canvas with objects, 26 × 26" surmounted by four tinted plaster faces in wood box with hinged front. Box, closed, $3\frac{3}{4}$ × 26 × $3\frac{1}{2}$". Box, open, $33\frac{5}{8}$ × 26 × 3". The Museum of Modern Art, New York, NY, U.S.A. Gift of Mr. and Mrs. Robert C. Scull. (8.1958). Digital Image ©The Museum of Modern Art/Licensed by Scala/Art Resource, NY. ART ©JASPER JOHNS/LICENSED BY VAGA, NEW YORK, NY.

There, your eyes would be confounded with four partial faces that are attached to the basic surface of the canvas.

Before trying to understand the confounding *Target with Four Faces*, we should detour for a moment to face what may seem an odd question. Are such combination works somewhat illegitimate as art because they are not *pure*? The word "pure" is used here somewhat in the sense of the Jewish dietary laws. Those dietetic rules require that *meat and dairy* products not be prepared, served or consumed together. Not long before Johns's work, there were many who followed similar artistic "dietary rules." Those decrees suggested that paintings themselves should be without any story and also be totally two-dimensional, this to isolate them from writing and three-dimensional sculpture. It was thought that each "medium has its own specialty ... literature is better at storytelling, sculpture can deal with forms in space, but painting is unbeatable at being purely optical, and flat."[2] These artistic dietary rules were practiced by some Abstract Impressionists with the blessing of critic Clement Greenberg.[3] Those artists stuck with flat paint, and stayed clear of adding solids. In many of his works, Johns conforms with an ultra-pure aspect of these artistic dietary laws: paintings should avoid even the *illusion* of a third dimension within them. However, some of Johns's creations (such as *Target with Four Faces*) could be called *impure*, as they include actual solid objects.

We humans— our artists included— are not only law-abiding creatures, but most of what we do, do with some thought, fits within one set or another of belief-based concept-carrying rules. In that sense, the Greenberg-given laws provided a valuable belief system for many. However, it is only fair to the reader that this writer disclose that such ideas are not part of his own belief system for art. As it happens, he belongs to the Church of Latter-Day Langer-Lovers. There, on the Sabbath the clergy provide devotional readings from the creed this philosopher (Susanne Langer) inscribed in good books such as *Feeling and Form: A Theory of Art*[4] and *Mind: An Essay on Human Feeling.*[5]

There are countless places to present arguments about creeds, but this isn't one of them. So let's move on now to a *key* to Johns's (perhaps ritually impure) art. We will see that many of this artist's creations display a singular vital idea. He is interested in more than the mere *objects* that are so prominent in his creations. He is also focused on human-like *subjects*, subjects in the particular defined sense that each subject has a "self or ego that causes something to happen." Subjects (in that sense of the word) are *agents*, "active powers or forces" with agendas to be accomplished. Such subjects *refer to* and "are concerned with" things other than themselves; they "direct their attention" to those objects. In this sense, subjects have (subjective) minds which they use to act with purpose in their object-full world.

Now we can look at how Johns uses that idea in *Target with Four Faces* and elsewhere. This work shows (above a flat five-ring target) a row of four three-dimensional casts of the lower half of a face. To a viewer, a human face — even a partial face — normally suggests an active *subject*, a human agent who is motivated to have his or her way with objects. But these faces (which lack the brain/mind part of the head) suggest humans as the opposite of subjects; they imply humans are mere objects.

Critic Leo Steinberg helps us with some related remarks on Johns's first one-man show, and on this work: "The thought of [some of Johns's works] gave me a distinct sense of threatening loss or privation. One in particular there was called *Target with Four Faces*.... I thought how the human face in this picture seemed desecrated, being brutally thingified — and not in any acceptable spirit of social protest, but gratuitously, at random.... And gradually something came through to me.... A dismembered face, multiplied, blinded, repeats four times.... It was as if the subjective consciousness, which alone can give meaning ... had ceased to exist."[6] Johns's truncated faces scare us with the suggestion that, fundamentally, we are (material) objects. Whatever else there is about us (that may be more than an object) is like a mere whistle on a train — which does nothing to move it along. Related mind/soul/spirit are nothing much more than "big talk." What is real for sure (in addition to natural things and human-made objects) are humans themselves as such basic objects.

The artist's *Device*, 1961–62 (Figure 26–2), deals with the same theme. It consists of flat paint plus two strips of wood, each nailed at one end to one side of the picture frame. Thus, *Device* looks like its wood strips can swing in arcs. Under each of the resultant half-circles, the paint is applied to look as if the stick had rotated and pushed it into position. That makes the sticks seem somewhat more than passive stick-objects. They seem like makers of semi-circular smears, lousy windshield wipers that spread gunk on the "glass" beneath them.

In *Device*, the three-dimensional wood strips seem like *agents* which act on the flat objects that are bits of paint. Significantly, that is just like the additional "dimension" of the subjectivity of humans that accounts for us being agents that act on objects in the world. At this point, someone might well ask: What about the self-governing machines we make that act independently as they roll on the surface of Mars or fly through the air on earth? Aren't they unlike fake or real windshield wipers? Don't they counter the argument that is being made? No, they don't really counter that contention; they heighten it. Such robot-like machines indeed seem to be

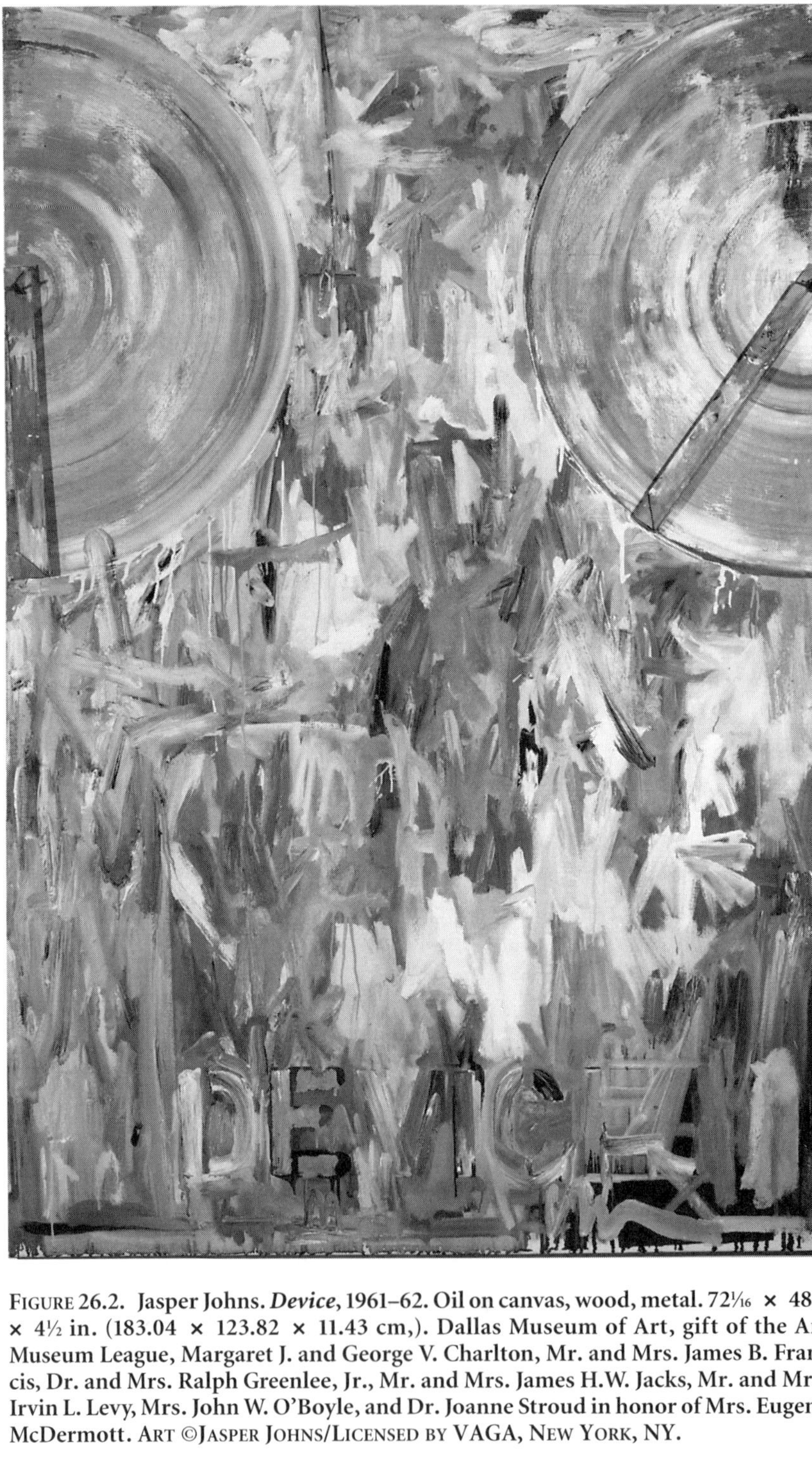

FIGURE 26.2. Jasper Johns. ***Device***, 1961–62. Oil on canvas, wood, metal. $72\frac{1}{16} \times 48\frac{3}{4} \times 4\frac{1}{2}$ in. (183.04 × 123.82 × 11.43 cm,). Dallas Museum of Art, gift of the Art Museum League, Margaret J. and George V. Charlton, Mr. and Mrs. James B. Francis, Dr. and Mrs. Ralph Greenlee, Jr., Mr. and Mrs. James H.W. Jacks, Mr. and Mrs. Irvin L. Levy, Mrs. John W. O'Boyle, and Dr. Joanne Stroud in honor of Mrs. Eugene McDermott. 

active agents like humans (or animals). But they increase our doubt about the importance of subjective feeling, spirit and soul. After all, they have no self or ego inside. Yet they act independently, even though there is nothing like a human or animal subject dwelling within. So, along with windshield wipers, these fancy tools only *seem* like subject/agents. More important, they make us seem to be no more than such objects.

In 1962, Johns made some works that further expand this idea. These disquieting pictures were made by pressing paper to a human subject's oiled skin, and then rubbing charcoal over the paper. They convey an eerie impression of a mostly invisible someone on the other side of the picture, some person whose body parts become visible only when pressed flat against the two-dimensional picture surface. For example, in *Study for Skin I*, 1962, what can be seen is the imprint of (an unseen human's) open hands and face pressing at the containing surface. Yet viewers cannot see the three-dimensional person himself. Doesn't that remind you that we cannot see the subjective human mind/soul/spirit that lives within the objective human body? One hard truth of subjective life is this: None of us has any *direct* evidence for subjective existence in others. After all, it is no more than a manner of speaking for someone to say, "I can feel your pain." In fact, it can't be done.

*High School Days*, 1964, is a work that refers to something beyond itself. It consists of a sculpted shoe with a small mirror stuck on its toe end. As suggested by its droll title, it takes an adolescent schoolboy — who is awash with new sex hormones — to conceive of sticking a mirror on his shoe tip. Why? Plainly, to let him look up the bottom of a girl's skirt.

Another work by Johns, although completely flat, repetitious and simple, is made movingly referential by its title, *The Dutch Wives*, 1975. This undramatic picture is split vertically into left-half and right-half panels, each with a plain checkerboard-like pattern of "tiled" squares. The right panel differs from the left one mainly in being a slightly deteriorated version of it. The creation's title makes it a work of buffoonery and bitterness. Writer Michael Crichton's book comments about the picture. "A Dutch wife is a board with a hole, used by sailors as a surrogate for a woman.... The painting seems to ask what the difference [between the left and right panels] is."[7]

We will see that the painting not only asks a puzzling question — it also answers it. The picture is an entertaining joke. It is also a joke with an acid bite. In part, it is a variation of the sex joke behind the expression "your turn in the barrel." The barrel joke is that men who go to sea are thought to get desperate and inventive about sex objects. When it's your turn in the barrel, you become an accessory to the "Dutch wife" hole in the barrel. (Spelling it out in plain words: The man in the barrel provides an orifice for his fellow male sailors to sexually penetrate through the barrel hole.) The vertical bar that splits this picture down the middle is the cross-sectional edge of the "barrel" or the "board." The left side of the picture represents the horny men subjects who seek sexual pleasure. The rundown right side — of this abstract painting of pornographic interaction — represents the men who are the Dutch-wife sexual objects who take their turn in the barrel.

The caustic aspect of the joke behind *The Dutch Wives* is unrelated to male–male sexual activity, but centrally related to Johns's theme about objects. Recall that there just isn't much difference between the left "subject" side and the right "object" side of this highly nonrepresentational porno picture. The implication is this: If your life is that of an inhabitant of meaningless, mindless flatland, being an active sex subject is only barely better than being a passive sex object.

To see one of the artist's lively — even *traumatic*— works, just look, look closely, at *Watchman*, 1964 (Figure 26–3). It shows someone who (wish though he may to stay a whole person) is quite literally falling apart and into the flatland. Occupying the upper right part of the canvas is a sculpted upside-down chair. Falling out of the chair and into the center of the picture is the three-dimensional leg and foot (but no more) of a seated person. In the picture's center is a confusion of broad brushstrokes that suggest countless streaks of dripping paint — as if the rest of the man had already fallen into and was part of the flat surface of paint. The picture's title instructs the attentive viewer: Watch man losing depth and disappearing as he falls flat (and short of what a person should be).

What we have in this artist's work is a world that is often devoid of real human subjects. It is a world chock-full of material objects— artifacts like the layered clothing, flush toilets, wheeled machines, and stratified buildings that we occupy during our daily life. They are all results of the objective knowledge that we have built up so rapidly and copiously that it sometimes overwhelms us. Such knowledge is in contrast with subjective knowledge, which includes what people have when they think about themselves and other people they care about, or the knowledge some may have when they think about how briefly and precariously they live, live with hopes of Heaven and God's love.

All of us human beings have always had both objective knowledge and subjective knowledge. Before the twentieth century, art's portraits of human life had objective and subjective knowledge entwined and not separated. What separated these threads of knowledge? Nothing did. They are one fabric, together for us now as they have always been. Then why have

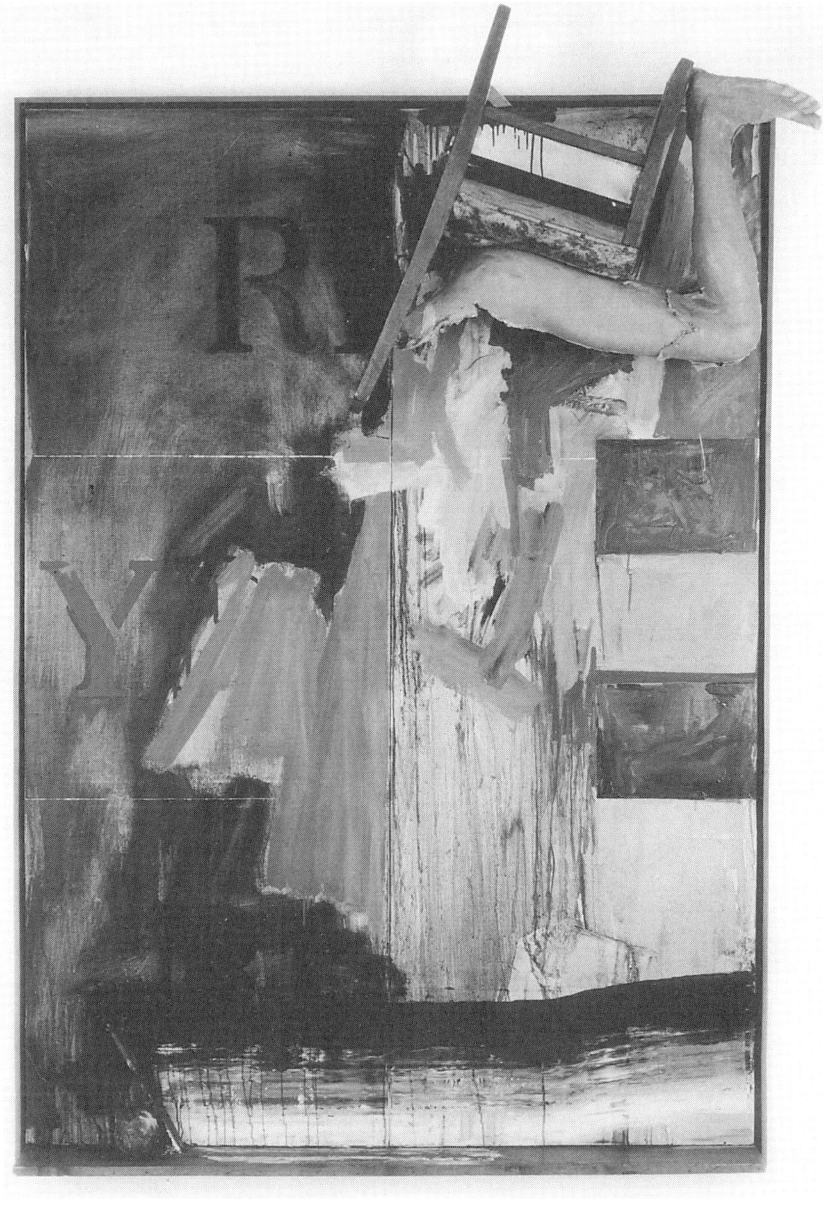


FIGURE 26.3. Jasper Johns. ***Watchman***, 1964. Oil on two canvas panels with objects. 85 × 60¼ in. The Eli and Edythe L. Broad Collection. Art ©Jasper Johns/Licensed by VAGA, New York, NY. PHOTOGRAPH COURTESY OF GAGOSIAN GALLERY, NEW YORK.

Johns's creations—which emphasize objects and deny or downplay the reality of subject-agents—become significant? Art makes a record for us of the import of living in the various eras and localities. In modern times and places, science and technology resulted in more and more new things. Medical science and technology developed knowledge of the things that are parts of the human body. This amounted to objective knowledge about ourselves as objects. We began to feel that the objective and the subjective domains were separate but unequal. Johns shows that we also increasingly started to fear that we ourselves may be mere bodies, and that to be no more than an object is difficult to endure.

## *Chapter 27*

# Art as Science Reporting: Robert Rauschenberg

Although there are people who find it bracing to jump into ice-cold waters, some don't do that. When they get wet, they do so little by little. That method will be extended here to Robert Rauschenberg's art by introducing it with two creations that are easier to penetrate than many of his others. After that we'll examine some of those others that initially seem uninviting. When we finally get into them, we will find them to be bracing.

Before taking a look at Rauschenberg's works, let's grasp the help of a guiding idea from critic Leo Steinberg. Of Rauschenberg's creations in which odds-and-ends such as newspapers, photographs and paint-marks are mixed with other things, Steinberg says they allude to "any receptor surface on which ... data is entered, on which information may be received ... whether coherently or in confusion...."[1] Something startling can be taken from that. Rauschenberg somehow makes art that—decidedly strange though it seems—is about what we normally associate with *science*: data and information, and their uncertain connection. This does not at all contradict the idea that art is mainly about the subjective side of life, and science is about the objective side. Rauschenberg's work is not science; it is art that shows some of our feelings regarding science.

Before we plunge into such alien waters, let's get into something more comfortable, *Monogram*, 1955–59 (Figure 27-1). Your eye will have no trouble figuring out where to look. Smack dab in the middle of the work is a goat, one with a tire around his middle. On the floor beneath the goat is a lot of stuff—paper, metal, wood, a rubber shoe-heel, a tennis ball, and

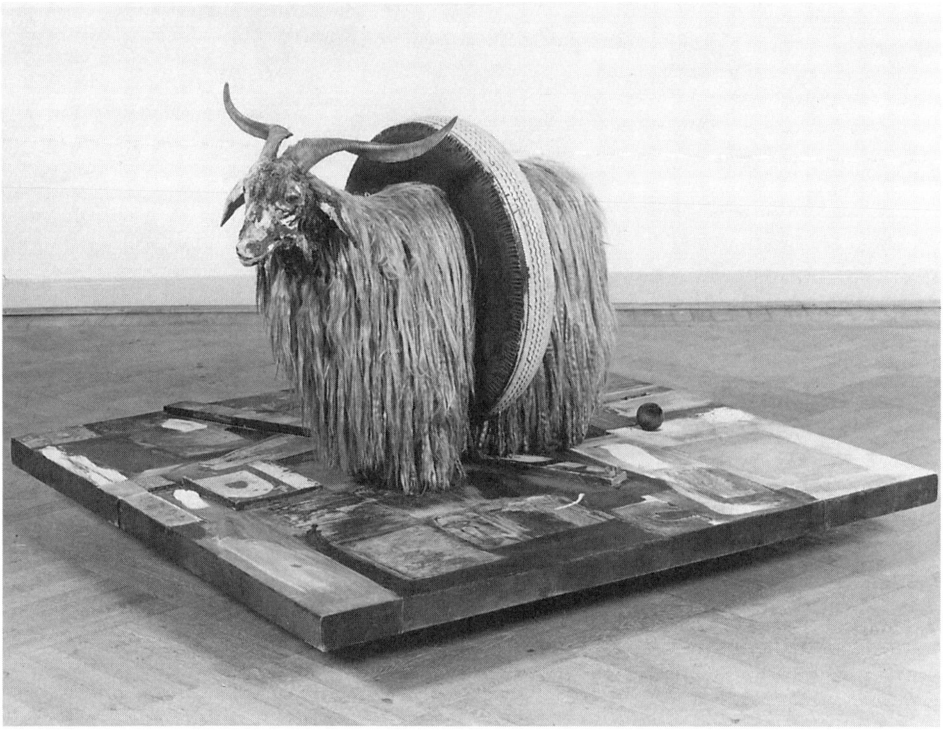

FIGURE 27.1. **Robert Rauschenberg. *Monogram*, 1955–59. Art ©Robert Rauschenberg/Licensed by VAGA, New York, NY. PHOTOGRAPH COURTESY OF MODERNA MUSÉET, STOCKHOLM.**

so on — which we will ignore for now. The tire-belted goat itself is fun, unexpected fun.

In his book about the artist, Sam Hunter tells us how *Monogram* came to be. "[In 1955 Rauschenberg] brought home a dingy stuffed Angora goat.... He shampooed its long fur, then daubed paint onto its damaged nose and placed a tire around its midriff. [When other uses for it didn't work,] he set it on a painted platform, its private pasture. [Critic Robert Hughes suggests the work became famous] 'partly because of its unacknowledged life as a powerful sexual fetish.'"[2] That may be so, but offered here is a broader explanation for *Monogram*'s grip on us. Its comic liveliness is what grabs us. The off-the-wall tire-belted goat on a private junk-filled pasture reminds us that life itself is sometimes no less than loony, nutty in the sense that is part of comedy.

One way of looking at human existence is to see it as including both (1) a tragic dimension and (2) a comic one. The tragic view sees a person's life as a single arc, initially ascending and then necessarily descending to oblivion. The comic view deals with the endless ups and downs within a life that goes on and on. "It is an image of human vitality holding its own in the world amid the surprises of unplanned coincidence."[3] The *Monogram* goat is doing what we ourselves sometimes seem to be doing — eating a path through a savory and unsavory mishmash.

A further step into the frivolous-seeming end of Rauschenberg's troubling waters brings us to *Sor Aqua (Venetian)*, 1973. In this creation, there are two things to look at: a bathtub and, suspended above it, some scrap metal with a piece of wood. Like *Monogram*, this work is off-the-wall. That's also what makes it lively, even though it has no once-living creature within the bathtub. It might well have been appreciated by Marcel Duchamp, who (by converting, a half century earlier, what had previously been a urinal into his *Fountain*) pushed ajar the bathroom and other doors to allow all kinds of weird objects to become art. In so doing, Duchamp opened art to the increasing importance of the endless human-made objects that came with science's astonishing expansion. Rauschenberg augments that theme.

So far, we have been wading. Another step or two now brings us to the deeper and more challenging part of Rauschenberg's work. We take them by considering two coupled works, *Factum I*, 1957, and the virtually identical *Factum II*, 1957 (Figure 27–2). The title word, "factum," means "a feat accomplished," which offhand doesn't seem helpful. *Factum II* is a number of things that have been brought to unlikely togetherness: The months and days of a calendar; a picture of two slender trees; two pairs of near-identical newspaper photographs; some painted patches here and there, some of it in seeming drips and drops; and yet other stuff. *Factum I*, except for a difference in the pattern of paint drips, is not distinguishable from its twin. Thinking about the two creations as a pair provides a bit of the comfort that we'd like but don't get from either work by itself. It gives us something in the way of an idea about them: Whatever they are separately, the two are almost alike. But what are they as separate works of art? As yet, that's too difficult to say.

The authors of *Modernism in Dispute* say something interesting about these two fact-filled works: "Rauschenberg makes a painting, *Factum I*, using apparently random juxtapositions and 'spontaneous' brushstrokes to produce an original configuration; but he then makes another version, virtually the same, *Factum II*, thus putting in question the very notions of randomness, spontaneity and originality."[4] That is tantalizing. If we were tempted to find that the "meaning" of *Factum I* is something about its randomness, spontaneity and originality, we really must give that up after looking at *Factum II*. Hey, we're not in a swimming pool, we just saw two hunks of opaque ice floating by.

Let's seek help again from Hunter's book about the artist. "[By 1954] Rauschenberg was no longer a minor vanguard artist.... His work had been repeatedly shown and elicited a storm of reaction.... Commentary ranged from charges that the paintings 'were beyond the artistic pale' to more sympathetic remarks like those of Dore Ashton, '[Rauschenberg] speaks of the nature of nature: Cloaked mystery and deterioration. Beauty is purity, he says, but decay is implicit. Appliquéd newspaper is his disdain of perpetuity. Life is cheap.'"[5]

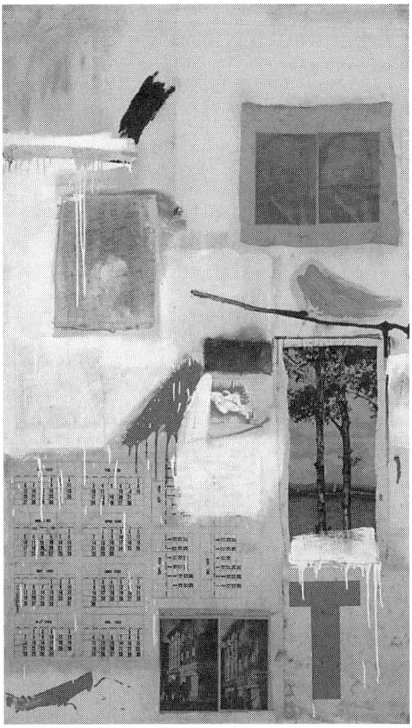

**FIGURE 27.2. Robert Rauschenberg (1925–) *Factum II*, 1957. Combine painting: oil, ink, pencil, crayon, paper, fabric, newspaper, printed reproductions, and printed paper on canvas, 61⅜ × 35½". Purchase and an anonymous gift and Louise Reinhardt Smith Bequest (both by exchange). The Museum of Modern Art, New York, NY, U.S.A. Digital Image ©The Museum of Modern Art/Licensed by Scala/Art Resource, NY. ART ©ROBERT RAUSCHENBERG/LICENSED BY VAGA, NEW YORK, NY.**

It helps us to learn that the artist speaks of the nature of *nature*. Let's modify that to suggest that his paintings are about the nature of the *objective* world that now is so full of human-made objects. Yet we still can't put together a clear picture of what's behind such works by Rauschenberg.

Here's more about what critics had to say about this artist. "Walter Hopps [wrote that Rauschenberg] 'sees and relishes the simple known facts of things, animate and inanimate....' Joshua Taylor [characterized] Rauschenberg's art ... as 'freeing the viewer to discover beauty and meaning where he might least expect it' ... a beauty that comes from order found, not order given, as if its permanent harmony existed precariously in a transient and unpredictable world."[6] We have just gained two valuable clues. Rauschenberg savors the *facts* about the world's things. Also, precious *order* in this world of facts is not given; it must be found by the interested observer.

Here (again from Hunter's book) are some remarks that confirm the importance of facts and order for this artist.

> Whether [Rauschenberg is] adding images by hand, transferring them with solvent and pen-nib or using silkscreens, novelty is a key objective. "There's the same quality of surprise and freshness that I have when using objects," he has said. "When I get the [silkscreens] back from the manufacturer, the images on them look different from the way they did in the original photographs, because of the change in scale, so that's one surprise right there.... Then, they look different again when I transfer them to the canvas, so there's another surprise. And they keep suggesting different things when they're juxtaposed with other images on canvas, so there's the same kind of interaction that goes on in the combines, and the possibilities of collaboration and discovery."[7]

None too soon, the fog above the waters is lifting. Rauschenberg himself confirms that he sought and found both *facts* about objects and an *order* that related them. He enjoyed putting facts and order together to make his artworks. Note that he did not do what most of us, who are not scientists, are tempted to do with the facts about and relationships between objects. He didn't complain about the world being too full of such tedious stuff.

Let's take what we have learned about these artworks and glue it together with some ideas about what these creations mean. Rauschenberg's works provide us with a stream of many new facts so that we may enjoy their novelty. Why do we need a whole stream of them? That's because, as it rapidly ages, each new fact "deteriorates" as its import decays. And why does the fizzle depart from the facts so quickly? Because that is in the very nature of facts. Facts are not where we humans are usually occupied; they are about the outside objective world. Facts, in truth, are not even *information*. The latter term refers to the *organized* (usefully put-together) facts that provide order for us. We put facts together in some order. We do that in hope of getting—you got it—objective knowledge that helps us make sense of the world and take advantage of it. Take one fact—say, that apples fall. Put it together with another fact—say, that the planets circle the sun. If you happen to be a brilliant scientist, that may set you on the lively road to understanding the generalization called *gravity*.

We needn't be patronizing by insisting on the obvious—that Rauschenberg himself is a skilled artist. The messenger is not to be confused with the message, the plain facts that he brings to us. He tells us that he views himself "as a journalist-artist in tune with ... the temper of the times. [He says,] 'I think of my activity more in relation to reporting than ... as something for an isolated elite.'"[8]

However, the editors and readers of the "journal" where his work gets "published" form a group called the art community. He knew very well what kinds of "stories" were appreciated and appropriate in the middle of the twentieth century. If he had reported the old kind of "art history" stories (such as the story of *Agamemnon sends his messengers to Achilles* with which this book's picture-stories started), they would have lost him his job before he found it. What are his stories about? They are about what is important in modern times: the artifacts and natural objects that fill our world, and how they are connected.

Rauschenberg makes beguiling stories about what we think of as the indifferent outside world. He shows us how there are objects in that world which, when put together, may at first seem like juxtaposed junk. However, we learn facts about them, and then organize that data into what science calls information.

Each of the pictures in Rauschenberg's twosome —*Factum I* with *Factum II*— shows us an orderly set of objects, something like two peculiar paintings, each of which shows a couple of objects: say, a falling apple and a circling planet. That there are two similar pictures emphasizes the connection between the objects within each. (In the case of falling apples and circling planets, the gravity of the connection was not known until recently.) However, each picture might show us a pair of things that allow a wrongly presumed connection, such as that between a *rotting* apple and a circling planet. We get no guarantees from science. Nor do we get any product-warrants from Rauschenberg's art that pictures the fact-full world of science. The uncertainty (about what, if any, meaning there is in such sets of facts) is one of the things that enlivens both science and this artist's work.

Rauschenberg offers us puzzling fun, something like the pleasure that scientists can get by trying to understand some aspect of the world. Look back at the goat-dominated *Monogram*. The junk on the floor makes the creature not only fun but also our kin in searching for something useful in a world full of trash. In doing his factum thing of "accomplishing feats," Rauschenberg brings us reports on the material world of natural and human-made objects that sometimes seem no more than higgledy-piggledy odds and ends. His light-on-feeling stories indulge in no lofty spiritual claims of a moral stench in human materialism. At the same time, his unemotional reporting simply ignores how the world has become more frightening because so many new artifacts are so dangerous.

# Chapter 28

# Art That Is Deeper Than Paper: Roy Lichtenstein

Starve your eyes on Roy Lichtenstein's *Whaam!*, 1963 (Figure 28–1). In a way, it's just as well that Matisse never got to see this artwork. Its total lack of decorative feeling for the inward side of life would have made poor Henri feel faint.

Lichtenstein went on from early Pop art paintings like *Whaam!* His technical mastery deepened as he pursued what — only at first — seems like the superficial and the shallow. Whatever his motives were for painting enlarged comic strip panels, the more useful question for us is this: Why would the art community embrace his work? The common wisdom is that the art world had its fill of the Abstract Expressionism that dominated most of the 1950s. The reaction formation took the shape of Pop art such as his.

Here's how critic Lawrence Alloway puts it:

> [Connecting art with high-minded meaning] reached a climax in the 1950s. The work of art was affiliated with a vague but exalted state, the immediate source of which was Abstract Expressionism. Lichtenstein and other artists of the subsequent generation ... resisted it; their problem was to produce art without metaphysical sanction.... Pop artists rejected the notion of a "tragic and timeless" art, in Rothko's phrase. They introduced current events and objects encoded in forms that were already flat: the flatness of Pop art therefore constitutes a shift in values away from the aestheticization and upward philosophical mobility of their predecessors. Lichtenstein's interest in discredited areas is a reaction against sentimental overestimations of the artist's role....[1]

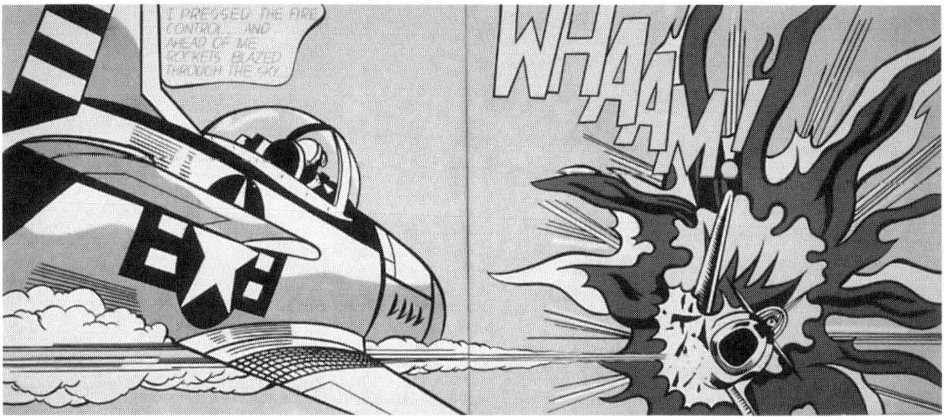


**FIGURE 28.1. Roy Lichtenstein (1923–1997) *Whaam!*, 1963. Magna acrylic and oil on canvas. 2 canvases, each 68" × 80" (172.7 × 203.2 cm.). ©Contemporary Art Resources, NY. Tate Gallery, London, Great Britain. ©Estate of Roy Lichtenstein. PHOTOGRAPH COURTESY OF TATE GALLERY, LONDON/ART RESOURCE, NY.**

Critic Christopher Knight expands on that: "Roy Lichtenstein was among a number of artists who playfully put the pop in Pop art between 1961 and 1965, sweeping away the pale doldrums into which.... Abstract Expressionist painting had by then so dismally fallen.... The reigning Expressionist faith in a self that can know only its own experiences, its own states of being, is given the boot.... [With] war-comic images of fighter planes, the battle over abstraction's stranglehold on contemporary art is addressed."[2] These words from Knight suggest that art went from lofty meaning to negligible meaning, from human glory or tragedy to trivia.

That may be true at an immediate level, but the barely bearable lightness of being that surfaces in Lichtenstein's work is not enough to account for it. Pictures such as *Whaam!* are not really light on significance, nor are they trivial. They are something heavier than airheaded fare. They are pictures about a world that has little or no meaning. To expand on that, try on this cozy metaphor: Art is a warm jacket. As you know, a jacket is only an insulator, an object with no intrinsic warmth. You warm it, and it prevents *your* continuing warmth from escaping to the cold world outside. To broaden this homely little homily, it is we humans who—by using them—bring warm-blooded felt life to all objects that enter our sphere.

Although the warmth and heat of sensation and emotion are part of subjective life, that inside life is also everything else we feel. That includes what seems to be the cold indifference of the objective world as it is sometimes represented in art. Lichtenstein and other Pop artists painted some of the cold lack of feeling that we *project* into the world of objects. We do that when the world seems altogether too much for us. (Objects in themselves don't have positive or negative qualities like "warmth" or "coldness." They have no such qualities at all.) But Popsters did their thing with a light hand, making fun of it and having fun with it. To be sure, there was a price paid for the fun. Pictures that seemed to trivialize meaning were eagerly acclaimed by an art community that sought to ignore the scariness of the real world. The price of downgrading meaning was to deny its central place in the human condition.

Lichtenstein's target was broader than Abstract Expressionism, so he could continue playing his game with other kinds of art from the past. Easier on the mind than his "comics" is *Stepping Out*, 1978. The woman there "is a Dalí-esque invention, consisting of lips, one blue eye, and a hank of hair that flows before a small head-sized mirror. The male is quoted from late Fernand Léger."[3] In *Stepping Out*, Lichtenstein has fun rolling his flattening steamroller over artists Dalí and Léger. He also continues to roll it over meaning.

Critic Arthur Danto tells us just how Lichtenstein did his flattening: "[He] uses the Ben Day dots of mechanical reproduction processes. So the canvases look like mechanical representations of vital gestures. But there is another level still, which we ascend to when we realize that the dots were not printed but painted in ... so we have artistic representations of mechanical processes.... What point could there be, for instance, to the dots, were someone unaware of the role dots play in mechanical preproduction and to the role of mechanical reproduction in the life of our culture?"[4]

Here, the word "mechanical" refers to the technology with which we produce endless human-made objects. Objects for consumption that come off of the assembly lines have to be used. Why? For one thing, because it's progress when we all get more objects. Television commercials, newspaper advertisements and junk mail are applications of objective knowledge that are put to use more commonly than is physics. They require a grasp of what really works to pop open our consuming mouths. And it is no accident that what works happens to be showing the viewer what seems to be the (subjective) pleasure of consuming.

Remember when art's change of life was beginning with Manet freeing himself from the need to conceal brushstrokes? For Manet it was a bold and risky gesture, showing his willingness to embrace the new openness that derived from science. Where now is that joyful liberation from the need to hide brushstrokes? If your retinas have regained their normal receptiveness, spend that sensitivity on Lichtenstein's *Brushstroke*, 1965 (Figure 28–2). Your overspent eyes will see a detailed picture of nothing

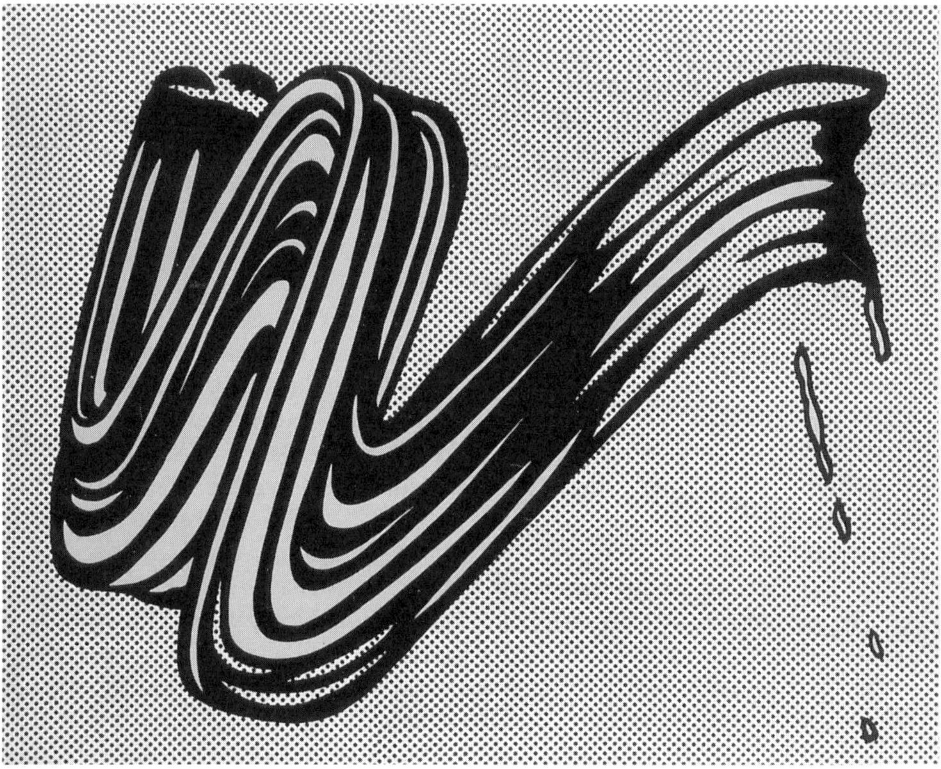

FIGURE 28.2. Roy Lichtenstein (1923–1997) ***Brushstroke***, **1965. Screenprint on paper. 56.5 × 72.4 cm. Tate Gallery, London, Great Britain. ©Estate of Roy Lichtenstein. PHOTOGRAPH COURTESY OF TATE GALLERY, LONDON/ART RESOURCE, NY.**

but the stroke of a large brush. Critic Dave Hickey suggests that, in postwar America, the use of brushstrokes in paintings to represent "spontaneity" and "authenticity" (construed as resistance to repression and regimentation) was no longer needed or appropriate. So Lichtenstein sought to achieve only a *graphic* effect.[5] Perhaps that is so, but Hickey's interest here seems to be mainly in the history of painting. As this book's interest here is in how art shows our stressed feelings about the strident twentieth-century changes in the way we live, that suggests a rather different kind of significance for *Brushstroke*.

Recall the picture of *Liberty Leading the People* (Figure 5–4) by Delacroix. Liberty's liberation has brought into being many things, including a bastard child. Its name, as your smarting brushstroked eyes tell you, is banality. That's somewhat like long-ago comedian Lenny Bruce's audacity as the first to publicly utter shocking words. Now, for the standup comics who use those once-unspeakable words in every sentence, they have become assemblies of sounds that serve only as *exclamation points.* Lichtenstein's brushstroke paintings are also exclamations, ones that orient the viewer only to brushstrokes. They pointedly insist that there is no longer much meaning in our world. His paintings seem so shallow that they suggest that our only choice in responding is whether to laugh or to cry.

Critic Alloway spells out how Lichtenstein made pictures like that. His "Brushstroke paintings have a rubbery consistency that implies high ridges of impasted paint, squelching out into occasional drips and spatters. The way in which Lichtenstein uses his meticulous technique to simulate a gestural process is ironic in the same way as cartoonists' explosions, a marriage of the quick and the slow."[6] There may indeed be something ironic about *Brushstroke*, but it is ironic in the sense of *paradoxical*, not in the sense of *satiric*. The paradox is how something as deadening as Lichtenstein's use of brushstrokes to paint nothing but brushstrokes can also be significant.

Both a clue to the paradox and a bit of comfort for your tired eyes can be found in the upper half of *Artist's Studio: "The Dance,"* 1974 (Figure 28–3). There you will find a remembrance of things past. Lichtenstein "quotes" from Matisse's 1905 *The Joy of Life* or from his *The Dance*, 1909. There, miraculously, a trace of the grace of the human dancers remains.

In the bottom half of the picture there is little sign of anything much. It amounts to a quote from countless past *still-life* pictures of fruit and bowls on tabletops. Those originals, to be sure, were both still and artistically alive with meaningful objects. But this bottom quote is not of happiness, sadness, or anything else that is alive; it is almost lifeless. Things in themselves are objective things, without life unless we project life into them. That is the paradox. What Lichtenstein does is to project *lifelessness* into his pictures. He shows us things that once seemed to pulse with life — now they are still.

Where is the life that late they led? It is, sob, totally dead. What happened to it? Did it just — Omygosh — die? If it did, that would be tragedy, but Lichtenstein paints no tragedy. He paints— well — next to nothing that seems of much central import. So his paintings suggest something deadly in a different way than is death. They hint that life itself didn't die; it was never alive!

Such an outrageously contradictory judgment should be clarified. And it will be, in a moment. For now, let's mix metaphors about art and clothing: Lichtenstein and other Pop artists tell us that the Emperor of Life is clothed in no warming jacket; he stands as a cold naked object.

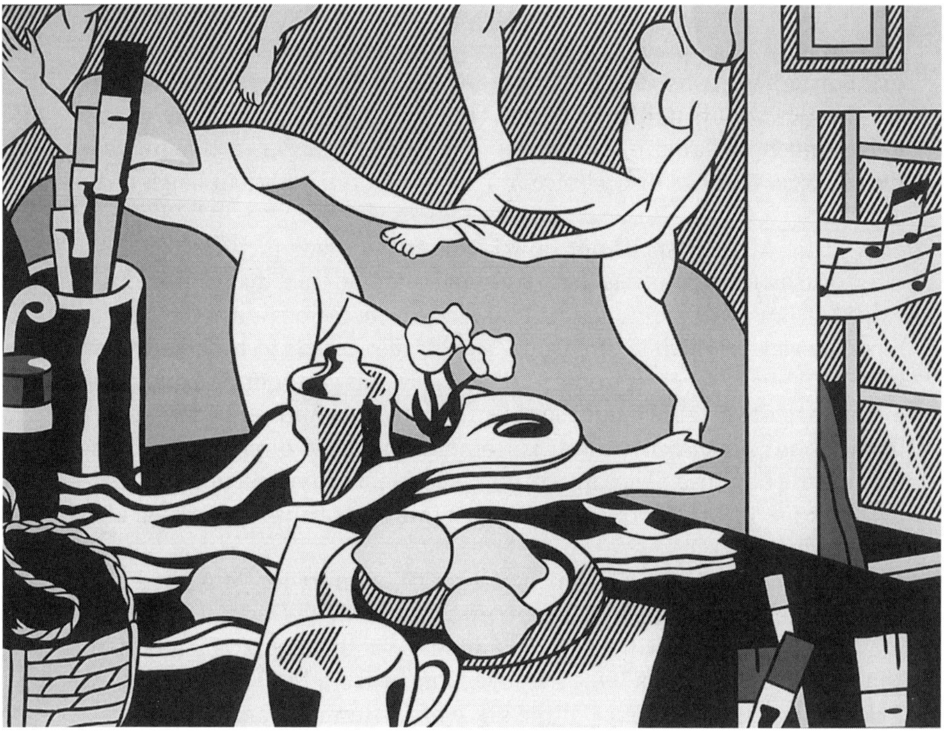

FIGURE 28.3. **Roy Lichtenstein (1923–1997) *Artist's Studio: "The Dance,"* 1974. Oil and synthetic polymer paint (magna) on canvas, 8' 1/8" × 10" 8 1/8" (244.3 × 325.5 cm.). Gift of Mr. and Mrs. S.I. Newhouse, Jr. (362.1990). The Museum of Modern Art, New York, NY, U.S.A. ©Estate of Roy Lichtenstein. DIGITAL IMAGE ©THE MUSEUM OF MODERN ART/LICENSED BY SCALA/ART RESOURCE, NY.**

Alloway writes: "Lichtenstein takes pleasure in a mask of simplicity. It is remarkable in retrospect the extent to which many critics took the dumb look of the paintings at face value.... Lichtenstein's apparent passivity seemed to be a denial of the personal element in art: his covert transformations ... were [not seen] at first."[7] The last statement — that this artist's work "seemed to be a denial" of subjective life — is not strong enough for the point needed here. Regardless of covert transformations, Lichtenstein's paintings *are* about the denial of feelingful life in art. By so strongly emphasizing the mechanically superficial, they show a denial of the personal element in human life itself. That is the sense — denying the subjective side of life — in which his paintings provide a view that life is without meaning. This (excluding here plants, protozoa, and other things that have no inner life) is a way of saying the worst: that life is *lifeless*.

Lichtenstein's topic is how human life seems to have become engulfed by—perhaps even part of—the objective world with its sometimes more harmful than helpful science, technology and endless artifacts. However, his art makes weightless trivia of that, as if to say: "Yeah, sure, there's stuff like these nuclear confrontations now and then. But do we worry? Hey, lemme tell you something. In this world, living with little or no feeling ain't half-bad!"

## *Chapter 29*

# The Emptier the Better: Andy Warhol

Andy Warhol was one of those rare artists who not only could paint their ideas but knew how to express them forcefully in words. In fact, he did even more than that; he adopted a public persona that matched his art and his words. Before we examine his message, let's consider some of his works.

Although he knew how to paint, talk about, and personify his ideas, he often needed help in getting them. When he was starting to make fine art in 1960, Warhol showed two related efforts to a friend. "One of them was a Coke bottle with Abstract Expressionist brush marks halfway up the side. The second one was just a stark, outlined Coke bottle in black and white. [The friend's response was that the first was crap], 'simply a little bit of everything. The other is remarkable — it's our society, it's who we are, it's absolutely beautiful and naked, and you ought to destroy the first one and show the other.'"[1]

There is a similar story behind *Roll of Bills*, 1962. In this money painting Warhol used pencil, crayon and felt-tip to plainly portray a roll of five- and ten-dollar bills. Late in 1961 he got talking with an acquaintance, Muriel Latow. "Andy said ... 'So, Muriel, you've got fabulous ideas. Can't you give me an idea?' Muriel said, 'What do you like more than anything else in the world? ... Money. You should paint pictures of money.... Either that or you've got to find something that's recognizable to almost everybody.... Something like a can of Campbell's soup....' Warhol himself was very open about the derivation of his works ... 'I was never embarrassed about asking someone, literally, what should I paint, because Pop comes from the outside, and how is asking someone for ideas any different from looking for them in a magazine?'"[2] With those words, Warhol gives us more than a mere hint about his work. He explained that he painted the objective *outside* and avoided the subjective *inside*.

Pictures of his work with Campbell's soup cans are not included here. In all likelihood, you know about it. If perchance you don't, it's worth noting that his first "work" with soup consisted of buying ordinary cans of it at the grocery store and bringing them home or to where they would be displayed. Later, among other soup works, he worked to represent each of Campbell's thirty-two varieties on a separate canvas. He also did real work on *Various Box Sculptures*, 1964 (Figure 29–1). For this creation, he made many facsimiles of cartons for scouring pads and tomato ketchup,

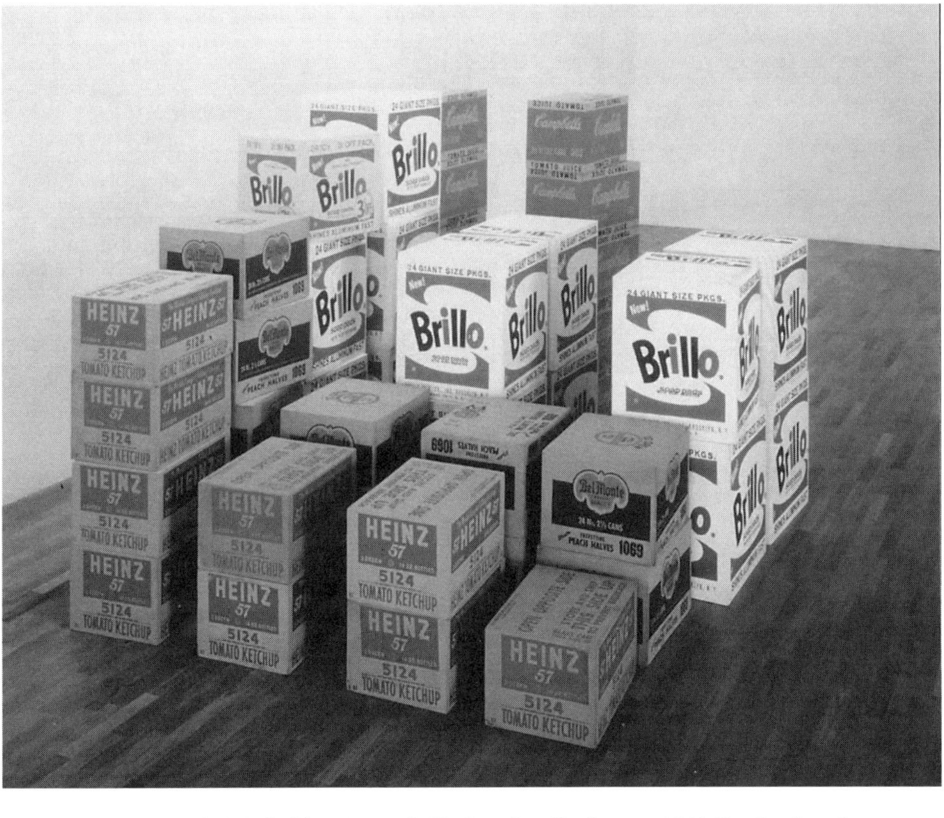


FIGURE 29.1. **Andy Warhol (1928–1987).** ***Various Box Sculptures*****, 1964. Synthetic polymer paint and silkscreen ink on wood. Dimensions variable. ©2004 Andy Warhol Foundation for the Visual Arts/ARS, New York.** PHOTOGRAPH COURTESY OF THE ANDY WARHOL FOUNDATION, INC./ART RESOURCE, NY.

and piled them up. Those facsimiles are so factual that he might have reduced the work by buying the real things.

A similar successful assault on significance is Warhol's *Thirty are better than one*, 1963 (Figure 29–2).This artwork basically consists of thirty tiny and plain copies of the famous Mona Lisa picture. Warhol put Mona through the flattening wringer by converting her to a bunch of mere *things*, objects depleted of lively subjectivity. Never knowing when or if he'd get another idea, he made maximum use and multiple works from Mona Lisa.

One of his many creations is based on someone (Marilyn Monroe) who was herself an outstanding performance artist: *Marilyn Diptych*, 1962. This artwork is much like the multiple tiny copies of Mona Lisa, except that its central figure is the famous performer. *Monroe's* artistic creation was to make herself the very embodiment of subjective sexual desirability. *Warhol's* work was to drain out every bit of that joyful juice with his images of her. In his book about the painter, Eric Shanes suggests that "Warhol heightened the garishness of the film actress's lipstick, eyeshadow and peroxide blond hair...."[3] Note that the word *garish* means "flamboyant" and also "tawdry." So we should sort out the two meanings. The persona that *Monroe* herself created was indeed flamboyant, whereas it was *Warhol* who reduced her to seeming tawdry and even insignificant.

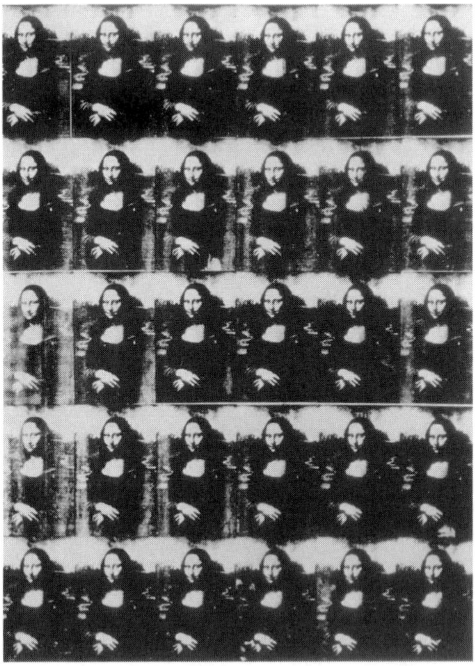

**FIGURE 29.2. Andy Warhol (1928–1987). *Thirty are better than one*, 1963. Synthetic polymer paint and silkscreen ink on canvas, 110 × 94½ in. ©2004 Andy Warhol Foundation for the Visual Arts/ARS, New York. PHOTOGRAPH COURTESY OF THE ANDY WARHOL FOUNDATION, INC./ART RESOURCE, NY.**

Here are more of the words that the articulate Warhol used about his work. "Writing on the topic of boredom, with which some of his work is associated, Warhol says that it is important to sit and watch the same thing—'I don't want it to be essentially the same, I want it to be exactly the same. Because the more you look at the same exact thing, the more the meaning goes away and the better and emptier you feel.'"[4] That is another bit of evidence from this open-minded artist. He painted several versions of many things—sometimes within a painting and sometimes with many paintings—in part so that the meaning in them would be better drained away.

In 1964, Warhol "announced to the press that painting bored him.'My real interest is movies, I make a movie every day....' His cinematic technique: mount the camera on a tripod, point it, let it run.... Lighting is harsh, film quality is poor, there is no sound."[5] There's no mistaking that Warhol's real interest in movies was in making *meaningless* ones.

Here's more from Shanes on that subject. "After 1963, Warhol began to feign an almost robotic emotional and intellectual vacancy.... His oft-quoted statement, first published in 1967, [was] that 'If you want to know all about Andy Warhol, just look at the surface of my paintings and films and me, and there I am. There's nothing behind it....' [Shanes then goes on:] But it is precisely in that very Nothingness that the artist's meanings reside, for it testifies to the meaninglessness that Warhol saw in the world around him, a negation that gives his art its acute relevance to the spiritual emptiness in which Western man for the most part lives."[6]

The end of the last statement—about Westerners living with spiritual emptiness—suggests that author Shanes may *himself* have gotten caught up in the same notions that Warhol's works illustrate. Elsewhere in his book, Shanes confirms this by writing: "[Warhol's] finest work ... confronts us with some of the underlying and most uncomfortable truths of our age, such as the unhumanity, exploitativeness, banality, trivialization and destructiveness of modern culture, as well as our loss of faith in God, art and life itself."[7]

A quarrel should be made not with Shanes in particular but with all of us when we start to think like that. The point is this: It is easy for any of us to get infected with the idea that there is *spiritual emptiness* in our culture, but that does not make the notion a *fact*. The truth is more complex. To be sure, Warhol knows just how to portray a world that is without much meaning. But it is our society's excitement about what is *presumed* to be our nothing-more-than-materialistic culture that makes us twitch and twang to Warhol's tuneless melody.

Here are a few similarly tainted comments from writers on Warhol's and other Popsters' work. "Remarkable—it's our society, it's who we are, it's absolutely beautiful and naked...."[8] "The vulgarity is shocking ... [Warhol's] images can be read as a pun on people, how much alike they are...."[9] "Pop art does not tell us what it feels like to be living [now]—it is merely part of [it.]"[10] "Instead of rejecting the deplorable and grotesque products of the modern commercial industrial world.... [Pop art has] turned to the wasteland...."[11]

This idea—that the Pop art of common objects shows us our vulgar, materialistic obsession with things— really misses the point. After all, we have no reason to believe that people before us—in any previous period in human history—were not just as interested in acquiring what we call "goods," which is to say things of whatever kind that were currently available. Likewise, it's doubtful that there has ever been a time—a time since goods cease to be utterly scarce—when folks didn't reproach themselves even as they welcomed fresh slices of the golden calf of goods. In the 1840s, for example, near the beginning of the modern times we have been examining in paintings, Charles Baudelaire wrote of "modern man suffering through an existence that placed material values above things of the spirit."[12]

The writers of *Investigating Modern Art* are right on when they suggest that in interpreting Warhol's work, we should "suspend the question of morality altogether."[13] Pop art makes little sense as a mere reminder and finger-pointer suggesting we are too materialistic. We get the same message endlessly at home, in church, and in the media. Our usual response to these messages is: "Isn't it awful? Some people are like that." Then, in a different mood on another day, "What's so bad about wanting a few nice things?"

So what else can we see in Pop art? Okay already, so creepy-looking Andy Warhol and others made Pop art of spiritless material, so what? First of all, here is something else. What Warhol's art, words and persona worry us with is not our disheartening materialism, but a different idea. Not materialism but *unfeeling material* may be all there is. It's not just front-men Warhol and other Pop artists. Behind them is the estranged spirit of the times, times wherein spirit itself seems to have disappeared.

Just how did this strange portrait of life come to be? It is the most recent of many changing responses of art to the growth of science and technology during the last couple of centuries. It is certainly not that science added little or nothing to human life. We know very well that it added a lot. We also know that nineteenth-century Impressionist art made upbeat pictures of a world that was increasingly based on reason and science. Likewise, Cubism sought to emulate the exciting innovations with which science and technology changed the world.

It wasn't until the twentieth century that the picture really got dark. Then technology's artifacts seemed not only to engage us but to engorge us. It was also then that science added enormous new terrors. At the same time, science subtracted something. It removed the confident assurance that humans are made of something more than cause-and-effect components that make up bodies. By default, it hinted that there may be no immaterial spirit within each of our bodies. That made us seem nothing more than material machinery.

Warhol got close to what's behind his art (but not close enough, as we shall see) when he said, "I think everybody should be a machine ... I want to be a machine...."[14] Warhol, with his vacant appearance and low-affect persona, was just the right guy to portray the spiritless emperor that we feared was our ruler. His empty-box sculptures are the right stuff to show the transparently trivial new clothes that we dreaded we had purchased with the cut-rate sale of our souls.

Warhol may have been well suited to expand mid-century anxiety about the meaningless machine nature of human beings. That does not mean we should take all that jazz as truth. It is, in fact, little more than trash-talk that replaces the truth. To start with, it was not true for Warhol himself. His pictures, words and persona (which presented emotional and intellectual vacancy) were "far removed from the sometimes overwrought and thinking [person] that he revealed to his friends in private."[15]

Warhol's own fear of feelings is of minor importance. What matters is the fearfulness within the society that made so much of his work. In the 1960s, when the art community and the larger society had such concern about a meaningless world of material, there was one large, if not overwhelming, feature of that real time and place. The 1960s were the years that followed the birth of hydrogen bombs. We in the U.S.A. and our enemies in the U.S.S.R. filled those years with the means to accomplish something new to humankind, mutual assured destruction with thousands of hydrogen bombs. That suggests that the concentration on too little meaning and too much material was a way to turn from the plague of too much meaning from too many bombs.

# PART VII

# Cracked but Not Broken: At the End of the Century

***Chronology of Technological and Scientific Developments***

1972: R. Leakey-Isaac expedition finds skull of early human, approximately 2 million years old

1974: U.S.S.R. space probe lands on Mars

1976: Spray-can gases found to damage ozone layer

1977: Neutron bomb tested, kills without harming artifacts

1982: RU-486, a pill that induces abortion, developed

1985: Antarctic survey confirms hole in ozone layer

1986: Chernobyl nuclear reactor explodes

1986: Space shuttle *Challenger* explodes one minute after launch

1990: Start of Human Genome Project

1990: Hubble space telescope launched

1995: First planet orbiting a far star found

1997: Lamb "Dolly" is first successful mammal clone

# Chapter 30

# New Faces, Old Spectrum

"Extremism in the defense of liberty is no vice.... Moderation in the pursuit of justice is no virtue." Those were the hard-to-resist portentous rhetorical words of 1964 presidential candidate Barry Goldwater. One of the things behind those noble-sounding words is the implication that, when the kitchen gets hot, the thing to do is to turn up the fire beneath the burners. Such ideas are often effective. Lyndon Johnson succeeded in fighting fire with fire by hinting that Goldwater's election might lead to a nuclear war. Extreme ideas are powerful because they seem to offer something invaluable, especially in times of troubled doubt: They offer certainty that the real truth is known.

To be sure, the extremist language of Goldwater and Johnson was in the political world of human power, whereas, in itself, art does not wield power but rather shows how people feel about aspects of their world. Nonetheless, in tough times, the desire to get away from doubt and uncertainty is common to both the political and the art communities.

We have seen that the content of art changed most dramatically near the middle of the twentieth century. First there was Abstract Expressionism, an intensified move toward portraying only subjective life. Following immediately was Pop art and related painting, an emphatic move toward the light-on-emotion objective end of the spectrum of felt life. These changes in art were in response to a mid-century America that had suffered two world wars and had embarked on what seemed the cold preliminaries of a third war that might destroy everything of value. Art first followed one aspect of complex social attitudes by turning inward, thereby avoiding the contemplation of such horrors. Then painting turned thinly outward, draining the feelings from the terror by portraying humans are spiritless objects.

The seemingly endless growth of science and technology certainly continued during the last decades of the twentieth century and the first years of the new one. Look, for example, at computers, which went from nowhere to everywhere in the last few decades. The author of *Heaven in a Chip* thinks that chip implants will soon make us—guess what—more like computers.[1] He also suggests that the distant leap is to replace the brain outright. Others who write about science propose that, before long, we humans will somehow move right into computers and live there—yes sir, live both better and endlessly. Yet others have a different view of our future with computers. "Nanobots," unbelievably tiny robots with computer brains—originally intended to be sent inside us to fix the flaws that arise in our bodies—may come by accident or cruel human purpose to do us great harm even as they themselves prosper. Although clones are not computers, they seem machine-like to us. Even if human cloning is banned in some places, rich and powerful egomaniacs may buy what they think will become multiple "copies" of themselves. Apart from such madness, genetic science may result in a world where we, as we are now, are no longer the ruling species, as suggested by a recent book, *Our Posthuman Future: Consequences of the Biotechnology Revolution.*[2]

Uncertainty about the nature and importance of our minds, spirits and souls continues to increase in reaction to the expansion of science. Does this suggest that objective knowledge will make subjective knowledge of mind/spirit/soul lose vitality and fade away? For sure, that "ain't necessarily so." In the United States and more so in Europe, religion—which certainly has a central place on the subjective side of our street—has suffered shrinkage. But it is not conflicting to say that it also enjoyed expansion. The (Mormon) Church of Jesus Christ of Latter-Day Saints, founded in 1830, has grown swiftly, especially recently in the twentieth century. Christian Science, created in 1866 with the bold adoption of the opposition's name, likewise expanded during the years that science continued developing so rapidly.

The second half of the twentieth century saw the rapid growth of New Age religious beliefs that often recognize no spiritual authority higher than personal inner experience. Such "Age of Aquarius" beliefs range from faith that God is fully within each human being to credence that crystals have healing powers, and that channeling to spiritual beings is real. Many forms of mental and spiritual yoga have become popular in the West in recent decades. Some scientists see evidence for the reality of a supernatural realm in brain activation patterns that show up when mystics and those who meditate do their spiritual thing.

The new widespread belief in the existence of UFOs filled with powerful unearthly creatures is not supported by evidence. A recent book[3] with the subtitle *Scientific Passions, Religious Beliefs, and the Pursuit of UFOs* suggests (among many other things) that some of us see UFOs as biblical end-time signs and wonders in the sky. Others see UFOs as satanic paranormal apparitions, or as phantoms intent on destroying the Christian faith. A broader interpretation is that UFO inhabitants are modern versions of both angels and demons. Likewise continuing to thrive are belief in ghosts, communication with the dead, and psychic healing. Also, good old-fashioned exorcism of invisible demons is "more readily available in the United States than perhaps ever before."[4]

Fundamentalist beliefs, although not dominant, are strong in our Jewish, Christian, and Moslem communities. Although being a fundamentalist hardly makes you a terrorist, extremist religious convictions surely played some part in the motivations behind the horrors of September 11, 2001. We have no prophet who can foretell whether or not the war on terrorism will be transmuted into what seems like a war between Western materialism and Middle Eastern spiritualism. We need no prophet, however, to tell us that Westerners and Middle Easterners are equally human. That suggests that Middle Eastern lack of Western material progress is more causally related to such a potential war than is a presumed Western deficiency of spiritual enlightenment.

If we turn back to our short-term history, we see the diminution and then disappearance of the Cold War with the Soviet Union. Society regained (before the emergence of terrorism) room for people to live with less thought of death from nuclear or other weapons of mass destruction. Recent decades have also seen the continuing advances of most sciences, including medical science. For one example, millions of third-world babies have been saved from dying of diarrhea. What saved them is oral rehydration therapy, which is simply drinking a mixture of sugar and salt dissolved in water, which prevents dehydration. The pre-terrorism decline of dire prospects for humankind was one factor in a move away from life-flattening Pop art over the last few decades. Desire for change may have been another. "As early as 1981, [critic] Hilton Kramer wrote that the turn to painting was a 'genuine change in taste....' What had been [regained] was the sensuous—and emotional—dimension in art.... Artists were desperate to reconnect with feeling."[5]

Mid-twentieth-century art had gone near the top (if not over it) with subjective and objective extremists. It is also true that near the century's end, despite the diminishing of mid-twentieth-century tough times, new art showed up at each end of the spectrum. That does not imply that this book's theory (about art's mirror cracking as it showed our two-way responses to the consequences of science) is incorrect. Rather, it implies that — like the devil who can quote scripture for his own purposes— artists (although hardly devilish) can use existing means for new purposes in an art world that no longer has clear limits. Art extremes apart, the return of the repressed middle ground also occurred near the century's end. New big-name "betweenist" artists found new ways of blending objective and subjective aspects of our knowledge in their pictures.

In the following chapters, we'll look at seven artists who worked during the last quarter of the twentieth century. Julian Schnabel creates mostly at the subjective end of the spectrum. In strong contrast, Damien Hirst jobs out all the details that make up his objective-end creations. The other five, Gerhard Richter, Anselm Kiefer, Eric Fischl, Jeff Koons, and David Salle, can be called betweenists who found pertinent ways to repopulate the middle ground.

## *Chapter 31*

# "Privately Maintained Realities": Julian Schnabel

Julian Schnabel's work provides company for that of the Abstract Impressionists at the subjective end of the art spectrum. However, his work provides no comfort to the true believers who think that two-dimensional art should not be contaminated by three-dimensional objects. Schnabel broke into the world of heavy-hitter artists by breaking pottery and putting the pieces into his painted pictures. Although he left the use of plate shards after that break-in, he continued his own way of making strongly subjective art.

It would not take an art expert to see what occupies the surface of Schnabel's eight-foot by ten-foot work, *The Death of Fashion*, 1978. It contains some paint and a lot of plates, mostly broken ones. Although the picture is loaded with such objects, it is a mostly subjective work. That means that — with the exception of its creator and those whose minds move like his does— no one knows much about what's beneath the surface.

So most of those who see an image of it don't know what the title tells them. One thing that is known is that behind *The Death of Fashion* is the continuation of an occasional fad that started long ago with the artworks of Kasimir Malevich. One of the latter, which carries the name *Black Square*, 1920, shows a black square. Another, called *Red Square*, shows you know what. These abstract paintings— which predate those of the Abstract Impressionists by decades— were believed to suggest a spiritual reality.

We might think that two such related paintings from Malevich (although perhaps one too many) are surely enough. After all, it does not take more than one or two such works for us to get the idea. Surprisingly, there is disagreement about that. Critic Arthur Danto writes, "The plain red square rendered me an exceptional philosophical service ... my appreciation of the differences between outwardly similar red squares ... set me on the path to art criticism."[1] Danto refers to both the idea and the actuality of a "museum of monochrome works ... a gallery of red squares ... [as being] of immense philosophical value."[2]

There should be no argument with the possibility that each of many black or red squares could have its own meaning to its painter. But surely the only justification for their acceptance by an art community would be based on the overly pure logic that—as each work equally meets the qualifications for art, they must all be accepted as art. Can you imagine anything more boring than a gallery of black squares? It's hard to believe that such "computer logic" would be acceptable to any real community of artists, curators, and critics. It is true, though, that there was once an exhibit in New York in 1966–67, called "The *Black Square* Room," of Ad Reinhardt's almost totally black square paintings.[3] However, they contained "Greek cross structures [which] slowly emerge against faint bluish or reddish 'haloes....'"[4] A search for such buried structures might have motivated viewers to proceed from one almost-black square to another.

A book (*Julian Schnabel*, 1982) from a gallery that displayed *The Death of Fashion* tells us that the painter "talks about the mediation of the 'pictures,' mental images if you wish, evoked by his experiences and the 'privately maintained realities' ... (specific to his own mind and personality) that he is able to manifest in the works.... [The use of the crockery] concentrates our attention on the very surface ... [Schnabel says] that one of the reasons for using the plates was to display the social usage of the materials and ask the audience to recognize that usage when unraveling the meaning of a work."[5]

Alright, let's try to understand the social usage of the crockery in *The Death of Fashion*. Here's a possible view. Schnabel couldn't previously achieve a deep penetration of the art world. The art community was like a donkey that ignored him. The way to get a domestic donkey to do what you want is—first—to get its attention. To do that, it is thought helpful to break a stick or two over its back or butt. Schnabel's audacious social usage of crockery was to show it broken in front of the art community. Although energy and nerve can serve artistic usages, if you've got them (which Schnabel clearly does) it's no demerit to use them to command attention.

The author of *Art of the Postmodern Era* expands on this usage: "The jagged shards of crockery provoked [a] kind of shock.... The broken crockery was special because it functioned as a counterpart to his impetuous brushwork. Not only did it provide a material and disruptive surface on which to paint, but it had metaphorical signification as well, suggesting violence, rage, chaos, and decay."[6] Wow! That sure sounds interesting, doesn't it? Who doesn't like a bit of violence, rage, chaos, and decay, as long as it's in pictures or the movies?

The 1982 book titled with the painter's name offers some more ideas about his plate pictures. "[He was] concerned with surface, with making a painting appear an object and with extending from that, through the use of the figure, to a commentary on the human condition.... Schnabel creates [this] with (what he describes as) an estrangement of the observer, an engagement created through pushing the viewer back to look more closely. Schnabel's figures ... make ... a violent avowal of inhumanity: the pain expressed there throws them back into the world.... Schnabel is taking the situations that are important to him and projecting them through his feelings...."[7]

Here's a summary of this view. The pain of human inhumanity prompted Schnabel to turn away from the objective world. He used plates to estrange the viewer through the strange broken surface of his work. This he did first to express that pain, then to fling it back to the cruel world in reproach. That makes a good story about *The Death of Fashion*. In fact, as a story (about a strongly subjective painting) there is only one problem with it. The story is in *words*, not something that the picture itself tells or shows. So it is likely that the picture does not convey that story to more than some of those who look at Schnabel's creation.

Although Schnabel dropped the use of crockery in his works, he continued to create paintings that were primarily subjective in their approach. One of them is *The End of Summer II*, 1990. The gallery catalog (which was published when this and other works of Schnabel were displayed) offers comments as follows:

> Schnabel has produced cohesive works imbued with a spiritual atmosphere that brings forth "the vibration of the soul" evoked by Kandinsky.... The four works in the series called *The End of the Summer* ... depict the mood of that time of year.... These colossal works, 16 by 16 feet, are painted on gray tarpaulin using oil and gesso [plaster of Paris, normally used to prepare a surface for painting]. They have a somber elegiac tone. Black and orange-red bands dominate [the first] three of the paintings ... Schnabel combines these controlled elements with others that are more accidental: amorphous white forms made with gesso.... A third element in these paintings provides a lyrical counterpoint ... Schnabel dipped wet linen rags in mineral-violet paint and flung them at the canvas.... Schnabel explains what the series means to him: "In *The End of the Summer* paintings, maybe I'm looking at the edge of the world, a place — heaven?"[8]

Another of Schnabel's large (this one is 98" × 79") subjective works is *Olatz mira* (Olatz, Look), 1991. "'The *Olatz* paintings are very Mediterranean,' says Schnabel, 'and very open.... They are not landscapes, but places that exist in my mind. I exist in those places.'"[9]

Although it's nice to see Schnabel's comment about how his work changes—from saying one work suggests human inhumanity to seeing others being portraits of heavenly or Mediterranean places—the comment remains what it cannot help being: only words.

What is to be made of such strongly subjective art as Schnabel's? It pretty much depends on the individual viewer to make his or her own sense of the work. Each of us has a private subjective world that is marginally knowable to others—knowable only by their ability to interpret the movements, gestures, words, and paintings that we use to represent (and sometimes also to misrepresent) our unseeable feelings and ideas.

If your soul is shaped like that of Schnabel's, you may be able to feel about his works much as he does. If what you see in his creations is nothing like what he sees, then, when your spirit meets his (perhaps as they wander about awaiting reincarnation in fresh flesh), yours may not recognize Schnabel's as kinfolk. How well such creations succeed with other people depends—to put it in counterfeit objective prose—on the modes of harmonic resonance shared by subjective natures. Remember (back in chapter 3) Richard Maurice Bucke's experience where he found himself "wrapped by a flame-colored cloud?" For him, it proved to be "Cosmic Consciousness" showing that the universe and God were wonderfully together as one. In this writer's view, Bucke had a highly unusual experience, one that resulted in a belief that the writer couldn't share. Recall the elder Henry James's sudden perception that some invisible demon in his room was sending out influences fatal to life. If this writer had an experience like that of James, he might twang and twitch in tune with James's notion that terrorizing demonic forces were behind it.

Most people find it easier to empathize with a hellish vision than with a heavenly one. For evidence of that, note that our writers with verbal artistry have given us more numerous, deeper, and more detailed visions of misery than they have of bliss. Were we desperate for a fake Darwinian adaptive reason for that, one could be put together.

For example, one such Darwinoid explanation might (by straining more than a bit) be based on the expression "A coward dies a thousand deaths, but a hero dies but once." The implications of that phrase go like this. When war comes, all the heroes and all the cowards sooner or later go to their deaths. Then, although the heroes stay dead, each of the cowards somehow returns from death and goes on to enjoy life, suffer war, and then die. Indeed, each coward does that nine hundred and ninety-nine more times. Thus those over-imaginative types live much longer than do their nobler fellows, which gives them time to generate a world full of their overly apprehensive kind of children.

Such nonsense apart, the formidable size and, perhaps, epic content of Schnabel's works may help explain why many unheroic types of people may not get much from his paintings. His work "'has a sort of romance of the ego and creativity.... He represents this operatic, Wagnerian ... power.'"[10] Tiny reproductions (not provided here) of his huge pictures might not be adequate help for you to judge for yourself the extent to which Schnabel's power is accompanied by something like the strength of Wagner's heroic music.

# Chapter 32

# "But Is It Art" Revisited: Damien Hirst

Damien Hirst is sort of an heir to Andy Warhol, who was somewhat of an inheritor of Marcel Duchamp's legacy. For evidence, sneak a quick peek at *A Thousand Years*, 1990 (Figure 32–1). Such a glance might not tell you just how Hirst follows Warhol and Duchamp, but you may not be ready for a real inspection of this raucous artwork.

The way in which Hirst succeeds Warhol and Duchamp is in taking or making objects that are accepted as art in much of the art community, and are thought outrageous in other places. All three make art that raise this related question: Why is *that* art? We have already explored the (in some ways artless) art of Duchamp and Warhol. To do the same with Hirst, we must summon the courage to look at his creations more closely.

More than a peek at *A Thousand Years* will reveal something outrageous: Maggots and flies enjoying the feast provided by the head of a dead cow. Hirst made "a sculpture that was two six-feet-square glass cases next to each other, connected by a circular hole.... But inside the vitrines ... there were flies and maggots ... and a severed cow's head. There was an ultraviolet light insect-o-cutor as well. There was a little natural life cycle going on.... The maggots hatching, the flies flying around and having fly sex and laying more eggs in the head, and the new flies doing the same, and all of them getting electrocuted...."[1]

Before trying to understand how all that is indeed art, let's get into something less disgusting (although more dangerous-seeming), a vitrine in which is suspended a tiger shark in formaldehyde. This work of Hirst's bears the name *The Physical Impossibility of Death in the Mind of Someone*

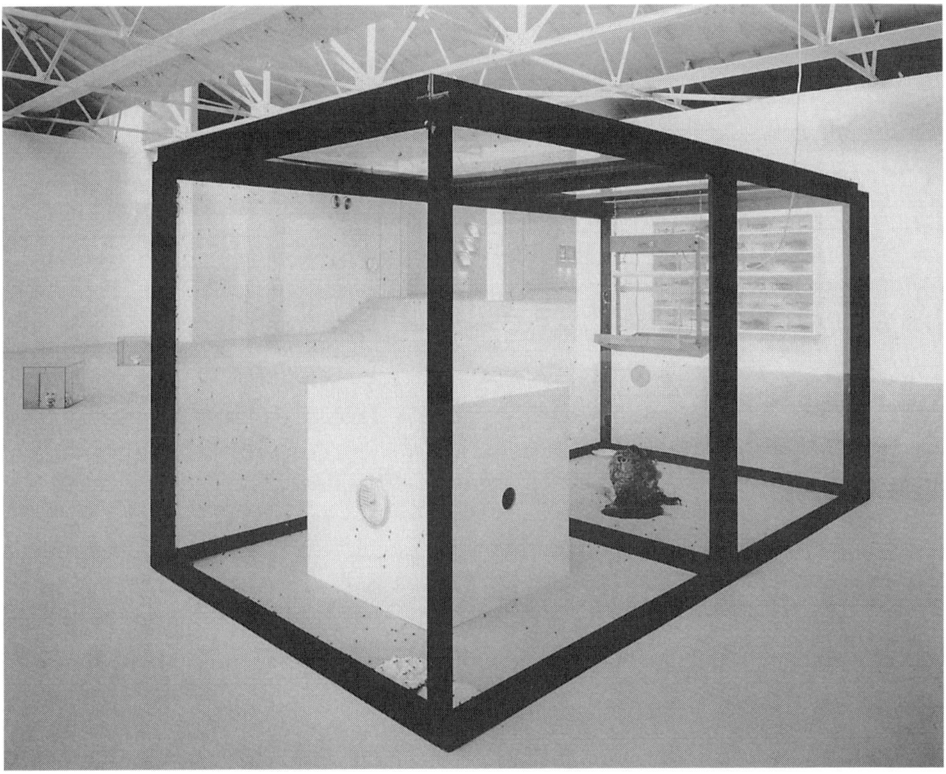

FIGURE 32.1. Damien Hirst. *A Thousand Years*, 1990. Glass, steel, MDF board, cow's head, fly zapper, bowls of sugar water. 84 × 168 × 84 in. (213.4 × 426.7 × 213.4 cm.) ©the artist. COURTESY OF JAY JOPLING/WHITE CUBE (LONDON).

*Living*, 1991 (Figure 32–2). In his introduction to Hirst's book, Gordon Burn tells us that after deciding to make a sculpture with that title, Hirst "lifted the phone and placed an order for an adult shark with a shark hunter who was unknown to him and who lived ... in southern Australia. When the commission had been completed, cash in advance, the shark was crated and air-freighted, unpacked and filled full of formalin ... then floated ... in a formalin solution in a tank...."[2]

Good grief. After all that you may need rest, so imagine yourself relaxing in a much more comfortable habitat, *The Acquired Inability to Escape*, 1991. It is a glass case "huge enough to have an executive style desk and chair inside it, with a lot of space round the objects, enough for a person to walk in and sit down, except that the vitrine [is] completely sealed. On the desk [is] a packet of ... cigarettes and an ashtray full of butts."[3]

All this stuff is defined here as art because it is accepted by most of

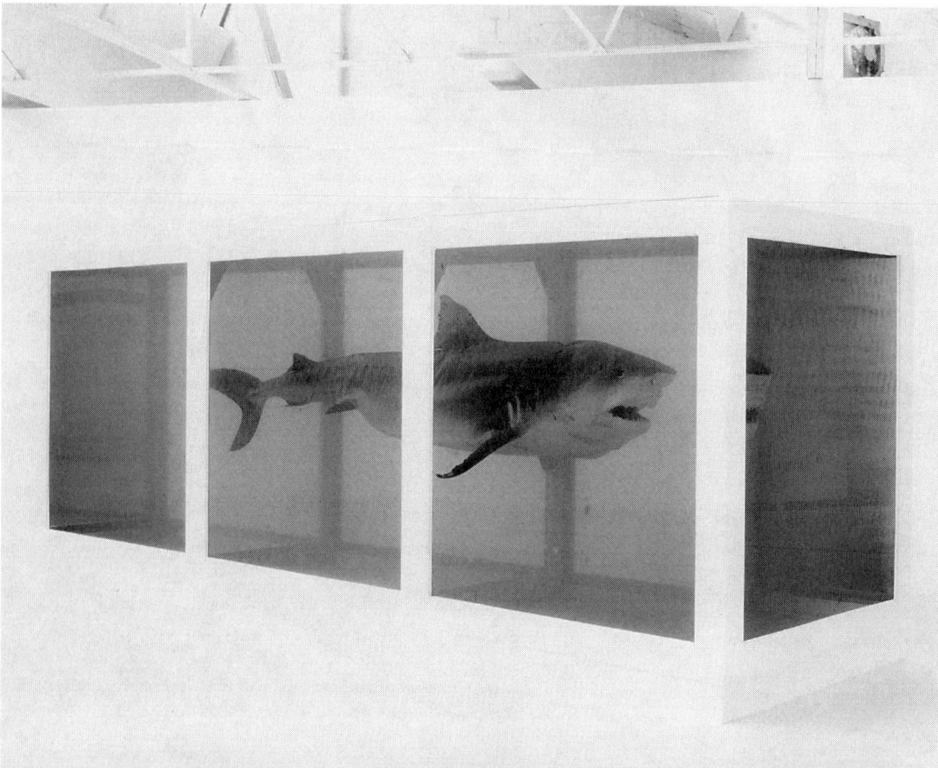

FIGURE 32.2. Damien Hirst. ***The Physical Impossibility of Death in the Mind of Someone Living*, 1991. Glass, steel, silicone, shark and 5% formaldehyde solution. 84 × 252 × 84 in. (213.4 × 1333.5 × 213.3 cm.) ©the artist.** COURTESY OF JAY JOPLING/ WHITE CUBE (LONDON).

the art community. These art-objects differ a bit from Duchamp's *Fountain* and Warhol's *Various Box Sculptures*. Duchamp's urinal and the models for Warhol's boxes were made for practical purposes. These filled vitrines were all made to (impractical) order, Hirst's order. More important, these glass enclosures full of one thing or another are similar to *Fountain* and *Various Box Sculptures* in this way. They return us to a variation on the is-it-art question: Why on earth does the art community regard them as *art*?

The answer is as it was with Duchamp's and Warhol's objects, as it has been since science began to make major new sense of — and countless additions to — the world's objects over a century ago. The answer is as it continues to be with scientists exploring the human genome with no intention of finding any chromosome segment that will invisibly flame out to become a soul or a spirit. Hirst's artifacts indirectly imply, they represent the idea, that what matters mainly for us everywhere is the objective world with its objects. Subjectivity is suspect and may not count for much if anything. And if subjectivity is suspect, souls and spirits are guilty of providing no direct evidence for their existence in other people.

Nonetheless, and paradoxically for sure, if you take or make an object and proclaim that it is art, you are attaching an intangible (subjective) idea to it. The object now *represents* something imperceptible, an *idea*. If the art community agrees with you, then you have made a work of art, even if the idea behind it cannot be put into straightforward words that can be widely agreed upon. In this case, the artworks that have been created are Hirst's various vitrines, many filled with rambunctious things.

It does indeed sound a bit like all an artist has to do is pull herself up by the bootstraps and—levitate. She has merely to wave a magic wand or say a few words which convert her work into art. Here is a caution of which artists are well aware. The triumphs of those like Duchamp, Warhol, and Hirst cannot easily be repeated. This kind of thing cannot be done every year. Decades have passed between the works of each pair of those three artists. During those years, the last previous version of this peculiar sort of art gets to be taken for granted. That allows something fresh to be valued. Also, you must be not just creative, but downright audacious, to get the attention of the art community.

Having fabricated a new niche of his own, Hirst went on filling it with more art in glass-walled cabinets. *Mother and Child Divided*, 1993, consists of a cow and a calf bisected lengthwise in four tanks. *Some Comfort Gained from the Acceptance of the Inherent Lies in Everything*, 1995, divides two cows into twelve chunks put into twelve tanks in random order. All of this suggests that Hirst's elaborate names for his objective art may amount to pseudo-titles. Perhaps all the vitrines have a common covert title. *This Object Is My Idea of a Work of Art. You Better Believe It!* Such a title would confirm our growing creed that the objective world is far more important than the subjective one. None of this, to be sure, takes away from the fact that Hirst has very much of a creative way of his own with objects.

Hirst also made flat artworks that have the same why-is-that-art character, such as *Arginine Decarboxylase*, 1994, which shows a nine-by-ten array of small colored circles or empty places. "He would draw out the circles with a pencil, in a simple grid, using a compass, and then fill them in more or less arbitrarily, with a different colour in each circle. Or get someone else to do it."[4] Clearly, this work is a well-organized pattern that is utterly low on emotional content. The pseudo-scientific pharmaceutical title contributes to the same implication of orientation toward the outside world. All of that puts it (with the vitrines) at the extreme *objective* end of the art spectrum.

Another work of Hirst's is titled *i want to spend the rest of my life everywhere, with everyone, one to one, always, forever, now*. This work consists of two-dimensional pieces of paper assembled into a three-dimensional work that is, in fact, a book. It is full of pictures and pop-up representations of his work. In the book, one picture is of a work of his that happens to bear the same title as the book. It shows "a ping-pong ball [bobbing] insouciantly on a fragile column of air, mimicking the artist's ... formidable ability to keep his balls simultaneously aloft and on the line."[5]

In this book of his, Hirst is quoted as saying, "I like the idea of a thing to describe a feeling. A shark is frightening, bigger than you are.... It looks alive when it's dead and dead when it's alive.... I hope at first glance it will look alive. It could have to do with the obsession with trying to make the dead live or the living live forever." Those words raise a question. Could it be that Hirst has provided instructions for his heirs, operating instructions telling them to pickle his body and preserve it in an antiseptic-filled glass-walled tank? Why? So that they could display it as a *Portrait of the Artist as a Dead Man*.

We have completed our visit with Damien Hirst. What remains, however, for us to consider is a challenging aspect of the question, "But Is It Art?" Switch your attention from Hirst's rowdy enclosures to something utterly different, the ripe paintings of someone by the name of Thomas Kinkade. Are you asking, "Who is this Thomas Kinkade?" You won't find his name in the 753 pages of *The Yale Dictionary of Art & Artists*[6] or the 384 pages of *The Thames and Hudson Dictionary of Art and Artists*.[7] Nor will you find a page with his name and one of his pictures among the 500 such pages in *The 20th Century Artbook*[8] or the 500 like pages in *Art USA*.[9] Kinkade is not listed in such books about art because — to put it bluntly — he is not accepted as an artist by the art community. "By and large, art critics consider Thomas Kinkade a commercial hack whose work is mawkish...."[10] (Worth mentioning here is another painter, Norman Rockwell, one who painted what became covers for a popular weekly magazine. Rockwell, listed in two of the four books above, straddles the fence of acceptance by the art community.)

Kinkade is brought to your attention here because some ten million people have bought his art-based products, such as lithographic reproductions (printed on textured-brushstroke canvas) of his paintings. Surely most of those millions made their acquisitions, at least in part, because they like to look at the pictures; they find viewing them to be a moving experience. That implies that, for those millions, each of Kinkade's reproductions that they buy has "the sensuous or poetic quality ... [that makes it the result of] a work of art. As such it may be good, [so-so,] or even bad, but it [has the right quality for art.]"[11]

That in turn suggests the obvious—the art community is an elite that disagrees with those millions of buyers. It finds Kinkade's works to be something else; by art community standards his work is just not art. So what? So nothing, except a reminder that the art community is an elite with standards that may not be popular. Kinkade, with plenty of popularity, is still hoping to get accepted by art's elite. (This writer should disclose that when he recently authored a book,[12] his own hopes were for both some popularity and some acceptance by an elite.) There are myriad elites who find that much work that relates to their special interests is just not good enough for them. They include barbecue-sauce elites, fish-catching elites, model-building elites, dance-performance elites, and low-fat (light-on-schmaltz) chicken-cooking elites. It happens that, for this writer, barbecue sauce from a good restaurant is good enough. He is indifferent to how fish are caught and to how models are built. When watching dance performances, he's usually for elite quality. As for cooked chicken, sometimes he finds a high-schmaltz piece of chicken irresistible. Likewise he finds that schmaltzy sentimental art, such as that of Thomas Kinkade, is sometimes fun. If you come across a piece of it in a nearby mall or on an Internet marketing site, give a look. You might like it.

## Chapter 33

# Striding on Two Levels: Gerhard Richter

As an artist, Gerhard Richter strides on two levels. In fact, he does that in two different ways. The first way is that he creates both (1) pictures that show something of his internal subjective life, somewhat like those of the Abstract Expressionists, and (2) pictures that are about the outside external world. Here, we will not look into his subjective works. Some of his pictures about the outer side of life will be more than enough of a challenge to occupy our attention.

These somewhat cryptic figurative pictures from this artist themselves show the second way he strides on two levels. They can be seen as showing (in the words of another artist, David Salle) that everything "in this world is simultaneously itself and a representation of [some] idea of itself.... [You see the thing itself] but you also see the thing ... [as your] ideas about it — as a representation."[1] We will puzzle over four examples of this two-tracked structure of Richter's figural works. Understanding them will require that we look beneath their surfaces. Doing that should give clues to the artist's vision of life in the world of the latter part of the twentieth century.

Let's start with one of his paintings that is easier to describe than it is to understand, *Window*, 1968. It portrays a window (which is parallel with the canvas) with white frames and clear glass panes in a four-by-six array. Through the pictured windowpanes, the painting shows the viewer the flat, bright surface of a wall that is close behind them. On that wall, what we see are the dark shadows of most of those window frames.

The picture's twofold nature is clearly that of (1) window frames and

(2) their shadows on the wall. However, what this artwork might mean is murky if not mysterious. Of help to us are some words of Richter's about another of his riddles about windows, one which is constructed rather than painted. It is called *4 Panes of Glass*, 1967. "[This creation consists of] four panes of transparent glass secured individually in black metallic frames and bolted together in a row between black metal rods.... In Richter's words, they allow one to 'see everything and grasp nothing.'"[2] These last words provide help, but not a lot of it. They imply that there is *something* we should grasp about these window creations.

The meaning of Richter's *Table*, 1962, looks to be beyond our grasp. In part, it is a painting of an ordinary white-topped folding table with plain black legs. However, it is painted so that it looks like the in-perspective picture of a table has been deliberately defaced by a hasty attempt to obscure the center of it with a few sweeps of a broad brush depositing paint on the canvas surface. Thus, *Table* is both distressed and disturbing to the viewer.

Surely it can't be that *Table* is no more than a confusing way of illustrating a certain old idea. That idea (one that preoccupied the art community for a century and a half) is this: What seem like three-dimensional objects in paintings are really no more than paint that was put on the canvas's surface. So, although unlikely, it is conceivable that the artist illustrates that idea by playing around with surface paint on top of the older paint that represented a three-dimensional table. Another simple explanation is also unlikely: Perhaps the artist disliked what he had first created, and defaced it in a moment of anger. And then chose to keep it for us to see and ponder over. The only thing about *Table* that we can be sure of is its two-aspect nature.

Richter's figurative paintings also include one that will help us grasp what is at issue with the above works. Its name is *Woman Descending the Staircase*, 1965 (Figure 33–1). Its central aspect — the descending woman — is lucid enough. It is the picture's other feature that is peculiar: Parts of the stairs— the ones that are behind the woman —can be seen *through* her and her gown.

Is the painter using the same idea here that Francis Bacon used with some of his pictures of a seated pope, where parts of a curtain *behind* the bishop of Rome are visible *through* him? For those pictures, the partial transparency of the pope may have represented the Holy Father's immaterial connections with the spiritual world. Here, with *Woman Descending...*, Richter might be lightly jesting or even smirking — suggesting that the soprano (Maria Callas) modeled in the picture sang like an ethereal, almost immaterial, angel even though she was indeed a corporeal, perhaps even carnal, human. (She was widely known to have a close relationship with a well-known, powerful rich man.) Note also that Richter chose a title that echoes the name of Marcel Duchamp's *Nude Descending a Staircase*. In Duchamp's artwork, the nude looks to be (as in time-lapse photography) in several places at once. That would allow some of the stairs to seem visible through her. Could it be that *Woman Descending* is merely a way to echo *Nude Descending*?

Of much-needed help to us is one of the simplest to see of Richter's pictures, *Betty*, 1988 (Figure 33–2). The subject in the painting (modeled by his daughter Betty), her blond hair above her bright red patterned jacket, has turned away from the picture's viewer to stare at an expanse of space that is, in fact, blank, totally empty of objects.

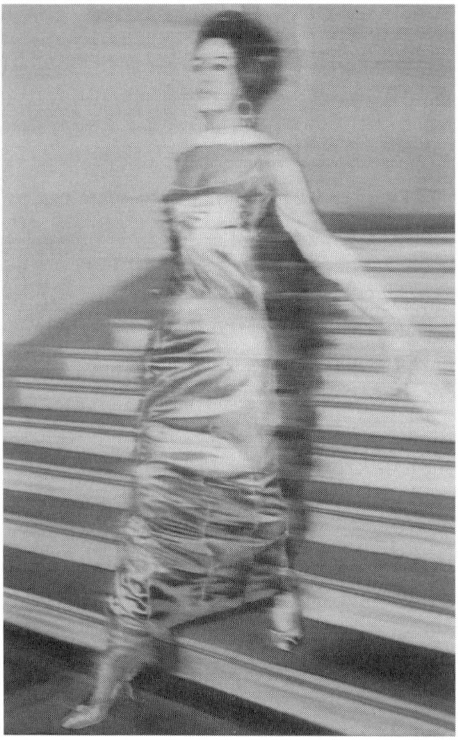

FIGURE 33.1. Richter Gerhard, German, b. 1932. ***Woman Descending the Staircase***, **1965. Oil on canvas. 200.7 × 129.5 cm. Roy J. and Frances R. Friedman Endowment, gift of Lannan Foundation, 1997.176. Photograph by Greg Williams.** IMAGE ©THE ART INSTITUTE OF CHICAGO. REPRODUCED WITH THE PERMISSION OF THE ARTIST.

The picture does not show us what is known about its making: When she posed, Betty was, in fact, looking at one of her father's many fuzzy, shades-of-gray figurative paintings. And indeed it would be a different picture if it did show that, as that might suggest to us only that the painter concentrates on his own work. So the viewer can only be mystified about what has captured Betty's interest.

Some comments about Richter together with certain of his own remarks will answer our calls (as viewers of these pictures) for "Help!" In discussing the artist, Michael Danoff notes that there are two "linked missions from which Richter has not swerved: investigating the varied visual models available for ... perceptions of reality ... [and] putting us in touch with [his] perceptions. [His] paintings are ... models of perception and [are also] ... depictions of reality.... From his earliest important works ... Richter has suggested that

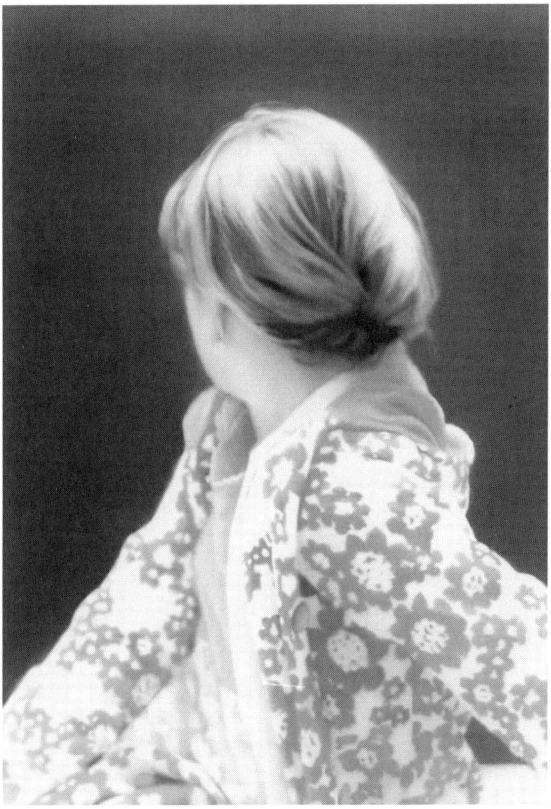

FIGURE 33.2. **Richter Gerhard, German, b. 1932. *Betty*, 1988. Oil on canvas. 40¼ × 28½ in. (101.2 × 72.4 cm.) The Saint Louis Art Museum. Funds given by Mr. and Mrs. R. Crosby Kemper Jr., through the Crosby Kemper Foundation, The Arthur and Helen Baer Charitable Foundation, Mr. and Mrs. Van-Lear Black III, Anabeth Calkins and John Weil, Mr. and Mrs. Gary Wolff, the Honorable and Mrs. Thomas F. Eagleton; Museum Purchase, Dr. and Mrs. Harold J. Joseph, and Mrs. Edward Mallinckrodt, by exchange.** REPRODUCED WITH THE PERMISSION OF THE ARTIST.

reality is elusive.... 'I cannot describe anything more clearly about reality than my own relation to reality. And this has always to do with haziness, insecurity, inconsistency, fragmentary performance, or what have you.'"[3] (The modesty of those last words of self-evaluation may be related to the painter's known dislike for ideologues who "know it all.")

Here are more of Richter's revealing comments. "I know nothing about the real, [about] reality. The only important thing ... is the translation."[4] "I don't mistrust reality, of which I know next to nothing. I mistrust the picture of reality conveyed to us by our senses, [a picture] which is imperfect and circumscribed."[5] The last quotation gets more focus if we insert *objective* in front of "reality" and *subjective* in front of "picture." Then the remarks would read like this: "I don't mistrust [objective] reality, of which I know next to nothing. I mistrust the [subjective] picture of reality conveyed to us by our senses, [a picture] which is imperfect and circumscribed."

We can now see that the artist focuses on these two things—the (objective) real world and (subjective) human "pictures" of it. That is in a way like philosopher-scientist René Descartes' words, "I think, therefore I am." In that phrase, "I think" can be viewed as making mental pictures of the world, and "I am" as existing as part of the world. Although Richter is not explicit about which comes before the other, Descartes' words may be reversed for this artist, with the *objective* part coming first: "I am, therefore I think." Priorities apart, we know that it is the combination of our senses and our *thinking* that gives each of us the basis for our mixed internal "pictures" of the reality outside.

Now we can see that in the above paintings, Richter is particularly interested in two linked things. One of them is *reality*, which is in good part unknown, elusive, and puzzling. The other is *our thinking that makes our pictures of reality*, which is like ideology in that it is insistent about what's what, it is often over-simplified, and it is sometimes incorrect. With this duality in mind, we can look again at the above four pictures and make some sense of them.

Richter's *Window* provides us with a variation on the old "Plato's Cave" story, a twist that reverses it. Remarkably, this plain picture of a window and its shadow succeeds in pulling inside-out the Greek philosopher's parable about shadows. Recall that the cave story (in Plato's *Republic*) is about reality and illusion. It tells us that what people in a cave see — on its walls — are only the shadows of the never-seen reality that is behind the viewer and outside the cave. Plato's allegory implies that, for mortals, the ultimate reality is unseen outside the world we inhabit. How then can people know that such an invisible reality exists? Only because some knowledge of it is somehow (perhaps by the grace of God, or by the good fortune of complex human genetic material) subjectively experienced, conceived and felt. Although we seem to interact with out-there objects, Plato suggests that all those things are like shadows; they are not truly real.

Richter's simple *Window* lets us grasp a story that is topsy-turvy to Plato's. Here, it is worldly objects — things called wooden frames and glass panes — that are *real*. What about the shadows of such frames? Shadows are not made of anything real; they are the absence of real light. But for humans who can think with their subjective minds, shadows are not merely negative but simplified pictures of real things. The shadows of the wooden frames are minimal *pictures of reality* that show no more than the shape of the things that they represent.

A meaningful story behind Richter's *Table* goes like this: Our picture of reality is so insistent that it obscures and interferes with the real world. The painted table in the picture represents *the real world* of objects, especially the twentieth-century human-made objects that we have most contact with. Most of the table can be clearly seen — that is congruent with the evidence that we can and do make our way quite well in the world of artifacts and natural objects. Note that objects, natural or human-made, make up a world that operates by basic laws of physical cause and effect.

The paint smeared on top of the center of the picture embodies a couple of aspects of *our picture of reality*. First, our perception of reality eclipses some of the reality beneath it. Thus, in Richter's picture, the paint that is slapped on top of the earlier paint *hides* part of the table painted underneath. The second aspect is that in our view of reality, we ourselves are actors at stage center. The brushmarks from slathering the paint over the table show the act of a human agent whose self-important activity is what matters most. Richter's *Table* shows that our picture of reality is a dramatic one with us acting out our will on the passive objects beneath our hands. He shows further that such a picture obscures the undramatic cause and effect that is central to interacting objects in the real world.

Let's move on now to seek an understanding of *Woman Descending the Staircase*. The artist may have seen a photograph of Maria Callas walking down some stairs. He then used it to show (1) an insistent (but not necessarily reliable) picture of reality, together with (2) some of reality itself. Here, the descending woman represents *our picture of reality*. The portrait of the woman in an evening gown is direct and pressing in showing her to be an attractive person. This impression hits the viewer in the eye, although it may or may not be true of the real woman who is represented. For example, we know that when she was young, Callas was much weightier, which would have made her a less attractive-seeming woman.

The stairs in Richter's picture represent *reality*. Parts of the stairs, shifted around a bit, shine right through the woman, showing some of the real stuff beneath our picture of her. Those stair-traces that show through her are like stardust, in that the fundamental heavy and light elements that make up the basic reality of our bodies (and everything else on earth) came from stars. For all that, what interests us most is not such stair-shaped stardust shining through the woman, but our own view of reality: here, the attractive woman descending.

Lastly, let's look again at *Betty*, and try one idea about it. As in the previous picture, the woman shown represents a view of reality. In this case, it is that of the artist. That view is one where people are constantly trying to make for themselves pictures of reality. The young woman is trying to make sense of the unknowable reality that is represented by the blank wall.

However, both reality itself and our views of reality are exceedingly complex things to deal with. So we could make a case that is opposite to the one just made. This opposite case goes like this. In contrast to the previous picture, here the woman in the center represents *reality*. The blank wall that she stares at represents a concealed *picture of reality*. The painting focuses its viewer on something we all take for granted: We each see the world — or at least the meaning behind it — in our own way. Betty sees things that we can't see. Likewise, there are things behind Betty that we perceive and she doesn't. The blank wall behind Betty is something upon which each of us projects a different private meaning.

One or the other of these ideas about this picture may be valid, but why does Richter bring up such things in a picture rather than in a classroom? If we regard the picture as being about something else the artist was concerned with when he painted it, this idea becomes less academic and more exasperating. For example, consider a picture that illustrates that "We can't see what a terrorist sees!" What the other person "sees" can become a matter, not of classroom debate, but quite possibly of life and death.

Richter is known for his aversion to ideology. In 1988, the artist painted not only *Betty*, but the series *October 18, 1977*, a suite of fifteen paintings based on photographs of the notorious German Baeder-Meinhof gang. They were regarded as terrorist-ideologists who believed that their world was again becoming like it had been under Hitler. These paintings from this artist are not conceptually black and white, but rather shades of gray. Many of them show the young women and men gang members in jail, some living, and some hanging dead or cut down from a noose. Of the series, Richter said: "All the pictures are dull, gray, mostly very blurred, diffuse.... I am not sure whether the pictures ask anything: they provoke contradictions through their hopelessness and isolation; their lack of partisanship."[6]

Chances are that the artist's daughter didn't share all of his views about the terrorist gang, and did not understand his paintings of gang members the same way that he did. With Betty focused on but unable to really perceive her father's picture, Richter may be showing his daughter looking at what he couldn't see, her picture of reality.

Gerhard Richter's pictures occupy all the modes in which painters have made paintings. Here we have looked only at some of his figurative art that shows that we all stride on two levels. One plane is that of the insufficiently known outside reality of the world we live in. The other is that of the — insistent, oversimplified, often unreliable, and sometimes just plain wrong — picture of reality we have inside us.

# Chapter 34

# Spirits in the Material World: Anselm Kiefer

Anselm Kiefer broke into the world of big-time art with an extremist move. He got attention by going from place to place in Germany with performance pieces wherein he gave the Nazi salute. "For examining the Nazi era ... he arranges a series of photographs for a publication.... His appearance in a series of staged tableaux ... attempted to shock and assault the taboos of German society.... 'I do not identify with Nero or Hitler,' Kiefer has said, 'but I have to reenact what they did just a little bit in order to understand the madness.'"[1]

With his painting, however, Kiefer is not an extremist given to overabundance in either the inward or the outward direction. He is a "betweenist" who (in the works discussed here) explores the large territory where the human spirit must live in a material realm stunned and warped by war. He also shows how immaterial spirit can be represented in a material world.

In *Emanation*, 1984–86 (Figure 34–1), we see one of the ways that Kiefer portrays spirit. Here, something other than paint is applied in a novel manner to show an emission coming down from the heavens above into the breaking waves of a seascape devoid of visible life. The emanation was "formed from poured lead [and] juts out several inches from the surface."[2] Perhaps the painter is suggesting that a new manifestation of Heaven's power must work its way into the sea before the waters bring forth life as we'd like it to be.

*Emanation* is an enormous canvas that is more than thirteen feet tall. The very size, oversize, of the canvas is a picture component that is used

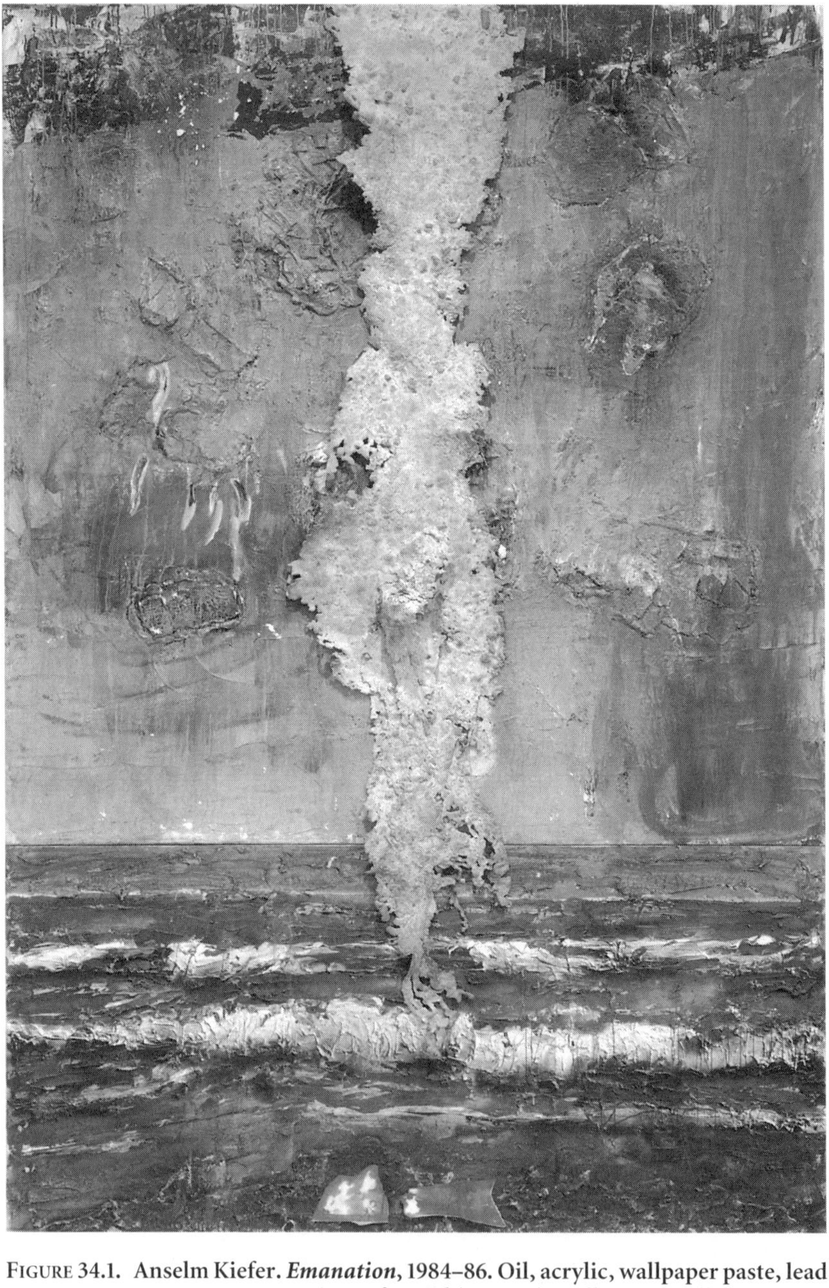

FIGURE 34.1. Anselm Kiefer. ***Emanation***, 1984–86. Oil, acrylic, wallpaper paste, lead on canvas. 161½ × 110¾ × 9¼ in. unframed (1990.92). Collection Walker Art Center, Minneapolis; Gift of Judy and Kenneth Dayton, 1990. REPRODUCED WITH THE PERMISSION OF THE ARTIST.

by several modern painters. In its essence, large size connotes great power. That view has come down to us through the millions of years when most of our genes were carried only by other species of animals. For most species, large size suggests great power. For example, some butterflies have the back of their spread-out wings patterned like two large eyes. Although it is only appearance, it may have kept predators away well enough to enable its success as a Darwinian adaptation.

For symbol-wielding humans, large size suggests both actual and symbolic power. We put the two varieties of that idea into our collection when we are small children under the control of large parents. For a painting, if the suggestion of power given by physical size is not convincingly backed up by the power within the picture, then the large canvas suggests propaganda more than it does persuasion. With only the small version of *Emanation* to look at here, the viewer may not be able to judge the success of its real size.

About ten by twenty-two feet is the size of a work of Kiefer's called *Deutschlands Geisteshelden* (Germany's Spiritual Heroes), 1973 (Figure 34–2). It shows a wooden room, a display hall with windows along the left and right walls. In front of each window, flames coming from urns are unambiguous in suggesting the title's long-dead spiritual heroes. Yet the darkened lower edge of the painting hints that the heroes themselves may now be scorched as a result of the wartime destruction of Hitler's Germany. In 1973, the artist created two different works with the same name, *Vater,*

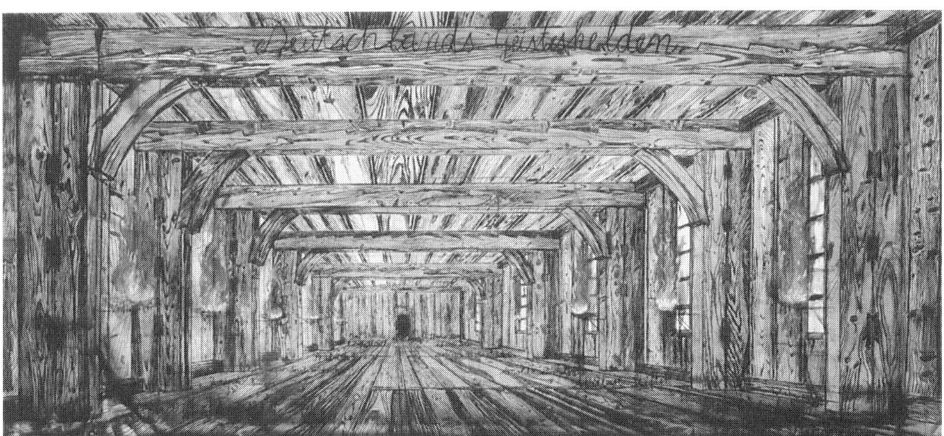


**Figure 34.2. Anselm Kiefer. *Deutschlands Geisteshelden*, 1973 (Germany's Spiritual Heroes). Oil and charcoal on burlap mounted on canvas. 120½ × 267¾ in. The Broad Art Foundation.** REPRODUCED WITH THE PERMISSION OF THE ARTIST.

*Sohn, heiliger Geist* (Father, Son, Holy Spirit). The smaller of the two is a rectangle of about five by five feet that shows the interior of a room, with three windows at the far end. The floor, side walls, and back wall (other than the windows) are made of plain wooden boards. In the middle of the room are three simple chairs, each in line with one of the windows behind it. On each chair is an urn, and coming from each urn is fire. The title of the creation suggests to us that the artist uses painted *flame* again, in this case to represent the three central spiritual beings of the Christian faith. This idea is reinforced when we note that the flames come from urns that are partially *transparent*, and that the urns sit on chairs that likewise can be *seen through*. These less-than-solid objects extend the idea that it is the spiritual realm that is represented in this artwork.

Representing spirit is a difficult problem for any artist. After all, spirit is normally regarded as different in kind from all objects in the material world. Luminous ever-changing fire is nonetheless a representation for spirit, one that may have been in our heads since we came up from the apes. The author of *Evolving the Mind* writes: "Consider a favourite analogy for things spiritual: a flame.... Cut off the gas [beneath a burner or] the blood supply to the brain and the flame or the consciousness disappears."[3] Although consciousness and spirit are not the same, they both seem immaterial to us, and they are intertwined.

The larger work with the name *Vater, Sohn, heiliger Geist* is shown in Figure 34–3. It consists of *two* rectangular pictures, one above the other. The top one is virtually the same as the smaller artwork with that name: A wooden room in which spirit-like flames come from see-through urns on transparent chairs. The rectangle below shows a rather different scene: a forest of bare-branched dead trees. This version of *Vater, Sohn, heiliger Geist* seems to show us two things. Above is the immaterial realm of spirit. The rectangle below, however, may represent the material German land that was scorched so badly under Hitler's rule. The juxtaposition of the two realms raises a question for the viewer: "Why are they shown together?" We cannot ignore the possibility that this artist is suggesting something challenging. Perhaps the scorching of the earthly realm and those who lived there *was the active doing* of the all-powerful heavenly spirits represented by the fire above.

Is Kiefer here joining those who use *negative* words rather than the neutral one, "mysterious," to describe their view of the way God works? Remember that for some — those who, long ago, believed in the radical Gnostic religious system — Jehovah (an early name for the Father in the picture) was conceived as being without love for humankind. With greater gods in a hierarchy above him, the Gnostics saw Jehovah as a low-level

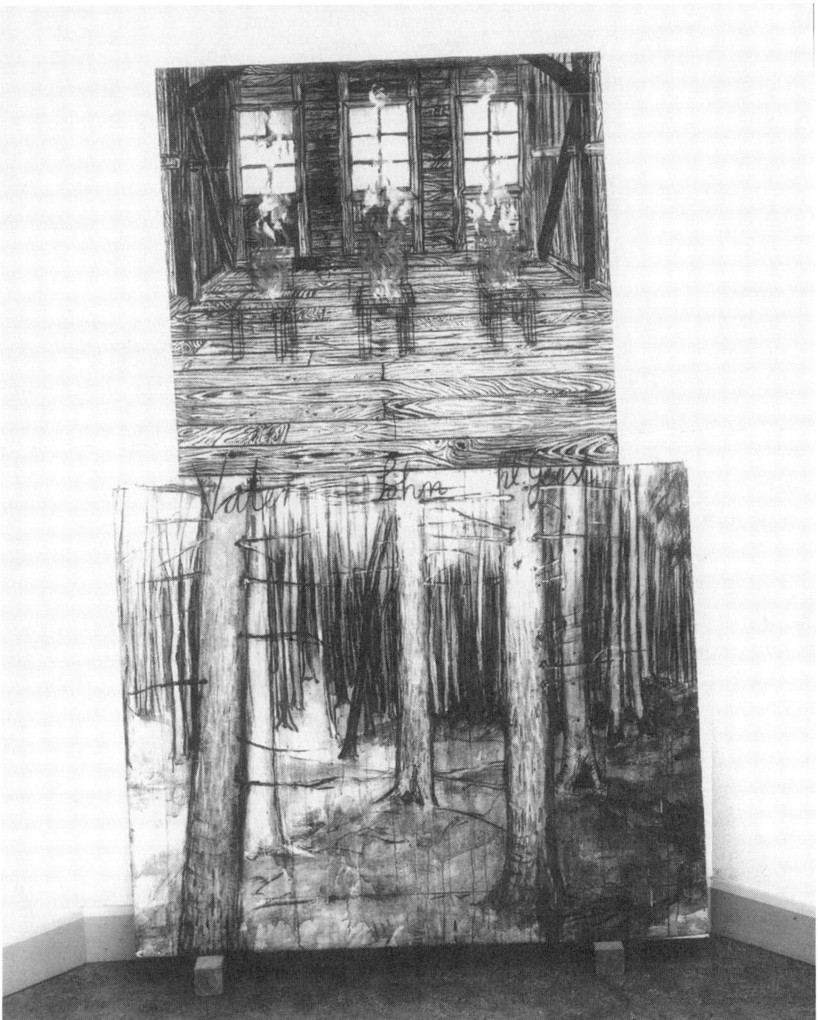


FIGURE 34.3. Anselm Kiefer. ***Vater, Sohn, heiliger Geist,*** **1973 (Father, Son, Holy Ghost). Oil, charchoal, synthetic resin on burlap. 290 × 189 cm.** IMAGE COURTESY OF THE ARTIST, USED BY PERMISSION.

deity who was not only powerful, but was the originator of darkness in human life. That extreme negative view of God is not common in modern times. On the other hand, there are two current views about God that do not seem to fit with this painting: (1) The view of a *remote* God, a benign active-mainly-long-ago First Cause or Intelligent Designer, and (2) the *near* view of a benign God that we now have access to as the once-fleshly Son and as the highly intimate Holy Spirit. So perhaps Kiefer is showing a certain *mixed* view of God, one that has long been associated with the Father, one formed before the Christian Son and Holy Spirit came to us.

That confounding view of a close patriarch God who can be both *lenient* and *severe* is illustrated for us within the Second Commandment: "The Lord shows forgiveness to those who love Him, and He visits darkness on those who do not." It is just such a combined view that can be seen behind one of our familiar sayings about war: "There are no atheists in foxholes." Whatever the unknown truth is about those six words from some believers, they present us with two ideas: (1) Non-believers are as much afraid of death as are believers. (2) When atheists face death, they find a reason to take up a *distinct* view of God. Their hope is that new, wholehearted belief will be helpful (but perhaps not sufficient) to induce (a surely authoritarian and perhaps egotistic) God to visit light upon them, rather than darkness.

Such a mixed view of the concept behind Kiefer's picture would suggest that the wartime devastation and death in Germany (which, after all, did leave most of the land and people free of destruction) may have indeed been a visitation of darkness upon many, a tribulation delivered (with little if any selectivity) by the first of the central heavenly spirits. But the mysterious picture itself does not seem to make that point. In fact, Kiefer does not give us any clues about the *nature* of the connection between Heaven's central spirits and the scorched earth's inhabitants. The cryptic larger picture called *Vater, Sohn, heiliger Geist* just shows us both the Blessed Trinity above and the devastated earth below. The artist's showing them together, however, certainly suggests that *there is a link* of some kind between the two.

Before leaving *Vater, Sohn, heiliger Geist*, note what may be obvious: We can't be at all sure just what this artwork (as well as many others) is about. The above comments on the picture provide not only *information* but also *opinions* about it. However, in a book about this artist and his work, we are given a rather different opinion about this artwork: "[Kiefer proposes that from his own] experience of the forest emerges thoughts about the existence of a divinity ... The process is thoroughly theoretical, hence the forest is an abstract one of rigidly horizontal branches.... [Three] chairs appear as symbols, but these finally go up in smoke.... The named subject of *Father, Son, Holy Spirit* demonstrates Kiefer's fundamental interest in the possibility of redemption. He is preoccupied with systems of thought that presume there is good in the world and offer hope if not salvation from the real evil that has existed throughout human history. Kiefer has said, 'I think a great deal about religion because science provides no answers.'"[4]

So then, is the picture about (1) the harsh Father visiting darkness upon a part of the wartime human world? Or is it about (2) the more loving Son and Holy Spirit providing hope of salvation from evil in our world? None of us can answer with any certainty. As the artist said, science may indeed provide no answer to questions like that, whereas religion (with the many different views inevitably associated with it) provides a plurality of answers.

Kiefer's *Adelaide — Ashes of My Heart*, 1990, might move the heart of a viewer more than do the more directly spirit-centered pictures described above. This creation shows no Adelaide. It simply shows her empty dress. But Kiefer does not stop by thus showing the ashes of life without her. He symbolizes the source of what may have happened to Adelaide by showing three toy bombers. For some of us, that may move the picture in a counterproductive direction. Consider one last piece from Kiefer, not a picture but a three-dimensional work, *Bruch der Gefässe* (Breaking of the Vessels), 1990. As simple as it is powerful, it shows — starkly and darkly — what were once containers for the work of the human spirit: charred and burned books, papers, and a bookcase.

Before we leave Kiefer, let's further consider his problem (and ours, as well) of representing immaterial spirit in a material world. Remember Delacroix's 1827 *Christ in the Garden of Olives*? In it, spiritual creatures, angels, were represented as humans with wings. Gauguin's 1889 picture on the same subject shows no clearly spiritual creatures (except for Jesus, and His spirituality is not shown but *inferred* from the title and the way Gauguin painted him). Kiefer's interest in spirituality harkens back to that of Gauguin. However, in *The Spirit of the Dead Watches* of 1892 Gauguin could still paint what looked like a strange and mysterious person and suggest (with help from the title) that it is a spiritual creature. Kiefer no longer has that as an option; he can no longer paint spirits as humans with wings or with spiritual expressions. He must use something else, such as solidified molten lead or painted fire, to represent the spiritual. This amounts to what could be called a *retreat*, not by Kiefer but by all of us. It is an acknowledgment of the utter materiality of the world, the overwhelming substantiality of objects.

For the most part, science has not sought to get an objective grip on the human spirit or soul. But it has been working on finding human consciousness in the material brain. Some scientists who work with the human brain regard consciousness as *identical* with a certain kind of brain activity. For example, in *Neural Darwinism*,[5] Edelman *identifies* immaterial-seeming consciousness with a certain kind of reverberation of neuronal signals between certain brain structures. The unexplained connection — as identical — of two such drastically different-seeming things leaves us unsatisfied. (An *unsatisfying* explanation does not imply an *erroneous or non-useful* one, as a glance at quantum mechanics will confirm.) Some recent work explores another possibility, that of a connection between a possible transcendental non-physical world and certain reverberating cycles in parts of the brains of those who are actively engaged in mystical practices or meditation. One book that explores this area is *Why God Won't Go Away: Brain Science and the Biology of Belief.*[6] Somewhat like Kiefer's material signs of spirit, certain cerebral echo-patterns can be viewed as somehow bearing witness to the existence of an immaterial realm.

## *Chapter 35*

# The Art of Storytelling: Eric Fischl

Many of Eric Fischl's pictures cannot be viewed as having stories that can be deciphered by the viewer; some seem to be intended to perplex. In them, the "sense that something is about to happen ... hangs silently but heavily in the atmosphere.... Always clear enough to be read, the activities in his paintings are perfectly legible yet utterly resistant to any attempts to get a confident grasp on what's really happening."[1]

Lots of them show people casually sitting in the sun on or near a beach wearing no clothing. One such is *Daddy's Girl*, 1984 (Figure 35–1), where we see a naked man in a deck chair who is holding an unclothed pre-adolescent girl in his arms. For a viewer, their nakedness, and also their vacantness, raises the question, "What's going on? Is what we see all there is?"

As elsewhere in both painting and life, the nakedness suggests sexuality. For *Daddy's Girl*, unanswered is this: Just whose salacious interest is exposed—is it the painter's? Is it that of the people in the picture—the daddy's or even the girl's? Or is it what the artist expects of the picture's prurient viewer? Note the glass of iced tea poised precipitously on the edge of the lower-right planter box. It shows us that the painter is well aware of the precarious situation he portrays.

The preceding question (about what more there is than naked relaxation) also raises larger inquiries implicit in Fischl's work: What, if anything, is there for these perhaps merely relaxed people, beyond pleasure and boredom? What about the life of the spirit? Is there something that saves them from a life that is ultimately empty? Do they do anything to

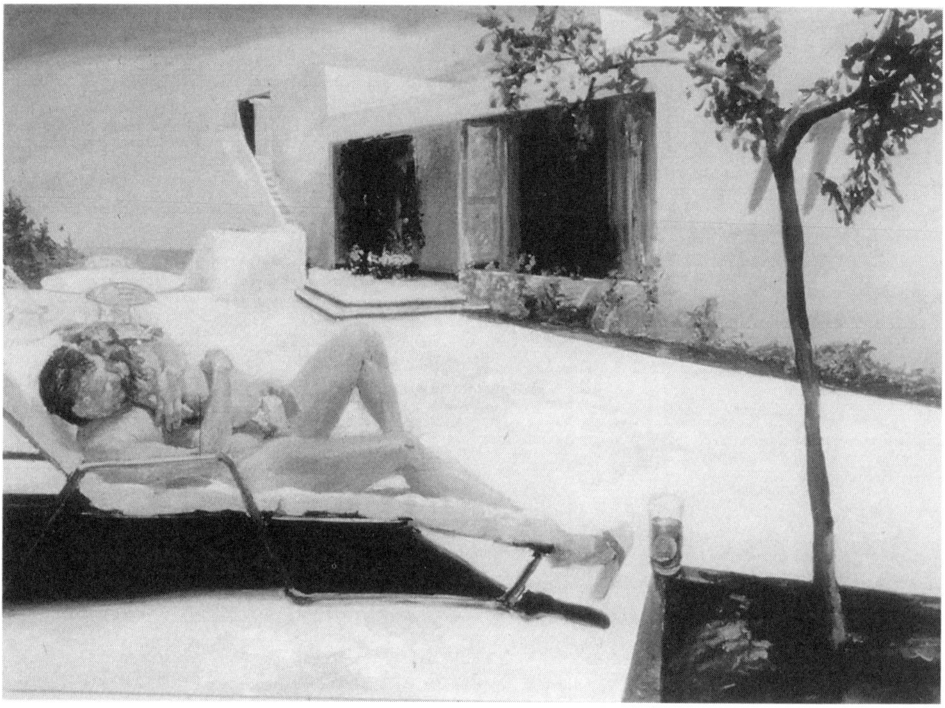

FIGURE 35.1. **Eric Fischl. *Daddy's Girl*, 1984. 78" × 108". Oil/canvas. Used by permission of the artist.** PHOTOGRAPH COURTESY OF MARY BOONE GALLERY, NEW YORK.

make life on earth better? For that matter, do they act on hopes for God's love and life eternal in Heaven? More generally, is this life of flesh and fabric all there is for them? Critic Peter Schjeldahl suggests that Fischl often paints a moralizing affirmative answer to something like the last question: The artist's "tragicomic images reflect the frittering away of American expansiveness, American economic and cultural swagger...."[2] That may well be so. Or it may not. It is hard to know how much a viewer reads into such pictures, and how much is really in them.

Many of Eric Fischl's pictures show us moments in people's lives that we might not expect. Some are so strange they are surely more imagined than real. For all that, many of these scenes of Fischl's are gripping pictures with stories, although what the tale tells is not at all clear. A glance at the artist's *Birthday Boy*, 1983, would show you two relaxed people on a bed, a boy and his mother, who are just calmly watching television. But the artist confronts us with their nakedness. The mother's spread-legged private parts are plainly within the close view of the boy, but he is not even sneaking a peak at them. Such a scene is certainly outside the experience, or even the imagination, of most viewers. Should we bet, then, that *Birthday Boy* is no more than an artistic variation on mass-produced risqué or erotic images?

Fischl takes the other side of that bet, and raises the stakes, with his *Bad Boy*. In *Art of the Postmodern Era*, the author relates that "*Bad Boy* tells the story of a boy-coming-of-age stealing from his mother — stealing a look at her vagina and money from her bag."[3] For those who see the picture, the Bad Boy certainly seems to be staring at his (apparently sleeping) naked mother's legs-apart crotch. That would put a viewer somewhere between turning away from something offensive and staying for a moment to peer pruriently with the boy. Fischl turns down the heat of the picture and limits the erotic tone by keeping her sexual parts in the dark. That in turn lets the viewer see that the bright-and-dark shuttered light is striping indifferently over her bed and her stripped body. As we will see, the woman herself might be (just like the Birthday Boy's mother) indifferent to the situation.

Not turning away from the picture would also give us the chance to see that the boy has his behind-his-back hands inside his mother's purse. Here, Fischl surely is not making a nudge-nudge wink-wink *Freudian* connection with where the boy's eyes are directed. That would be little more than a cheap shot. So what exactly is the boy doing? And, for that matter, what is the painter doing with this picture?

Perhaps the artist is making what we used to see in pictures, an illustration of a *story*. At first blush, such a striking but straightforward story might go like this: The son came into the room to pilfer his sleeping mamma's purse, not to peek at her privates. Once he's there, he can't keep himself from staring where he knows he shouldn't.

Let us imagine further events of such a story. If Mamma wakes suddenly and finds him there, she might metamorphose in a moment from a conflict-inducing sex object into a fierce female or an ogre who raises hell with him. The beginning of such a story is what the picture seems to show. Look, however, at some contrary remarks by the painter himself about the woman in the picture. "Hot, idle, a little bored, she's picking at her toenail. Look at what she's showing us!"[4]

Wait a minute! If she's idly picking her toenail, then she certainly is *not* asleep. That suggests the woman doesn't care that the boy is looking at her private parts. Doesn't that make her a Bad Mother? What then about the boy? Maybe he's seen her vulva more times than he can say. Maybe the boy is merely making sure that his mother doesn't notice him stealing from her purse. Is he a bad boy only because he sneaked in to take her money?

Wait another minute before accepting that version of the story. The above words from the painter go on to say that the boy "can't take his eyes off her." Does that mean the son *is* staring at what he shouldn't? Or instead that he's only watching to see that she doesn't catch him at what *he* is doing? The truth of the matter is that we can't be sure what the picture is about. Further, shouldn't the picture speak for itself? Doesn't needing help from the painter's words about it suggest that he hasn't succeeded with the picture itself? The plot is thickening.

Let's stir up things further by bringing in something critic Peter Schjeldahl says about the painter's childhood with his own mother. "Fischl's mother [was an alcoholic who] ran off the road in the car she was driving alone and was killed.... The worst part of having an alcoholic or otherwise addicted parent is probably the realization, unbearable for a child, of being loved less than a mere substance...."[5]

This permits a different kind of interpretation of *Bad Boy*. We know that a young child will often refuse to believe Mother is uncaring, and will prefer to bear the blame for whatever the situation is. So the nail-picking pictured mother might not really give a damn about her son or anyone else. Perhaps the adult painter who was once a boy can't quite bring himself to paint a Bad Mother. It's easier to blame the boy, although his transgression is far less important than her unbearable indifference to him. Or maybe Fischl — by showing a detached mother and a boy whose behavior is not really very bad — is purging himself of some of the guilt he wrongly assumed in his childhood, and also coming to an acceptance of his mother.

Only this much is clear. Many nineteenth-century painters, such as Ingres with his paintings about Achilles and Oedipus, *illustrated* previously well-known stories. Making up a new story within a picture makes for a more confusing result, as it is not easily done without ambiguity. If nothing else, it shows us that for stories to be really unambiguous, what's needed are *words*. Although pictures are more moving than ordinary discourse, they don't move us to a clear common conclusion.

Fischl's *Once Where We Looked to Put Down Our Dead*, 1996 (Figure 35–2), deals explicitly with death (and tangentially with sexuality). The picture shows— each bare of any cover of clothing — a man carrying a young lifeless woman. The two are illuminated by a shaft of light where the man has paused for a moment in the dark room. What is this place where he hesitates? Is it a mausoleum for those who have left life? Is it a museum that shows what the once quick were like? Although the picture does not spell it out, Fischl may well be illuminating something that is much more important. He shows us that the woman's man, wherever he

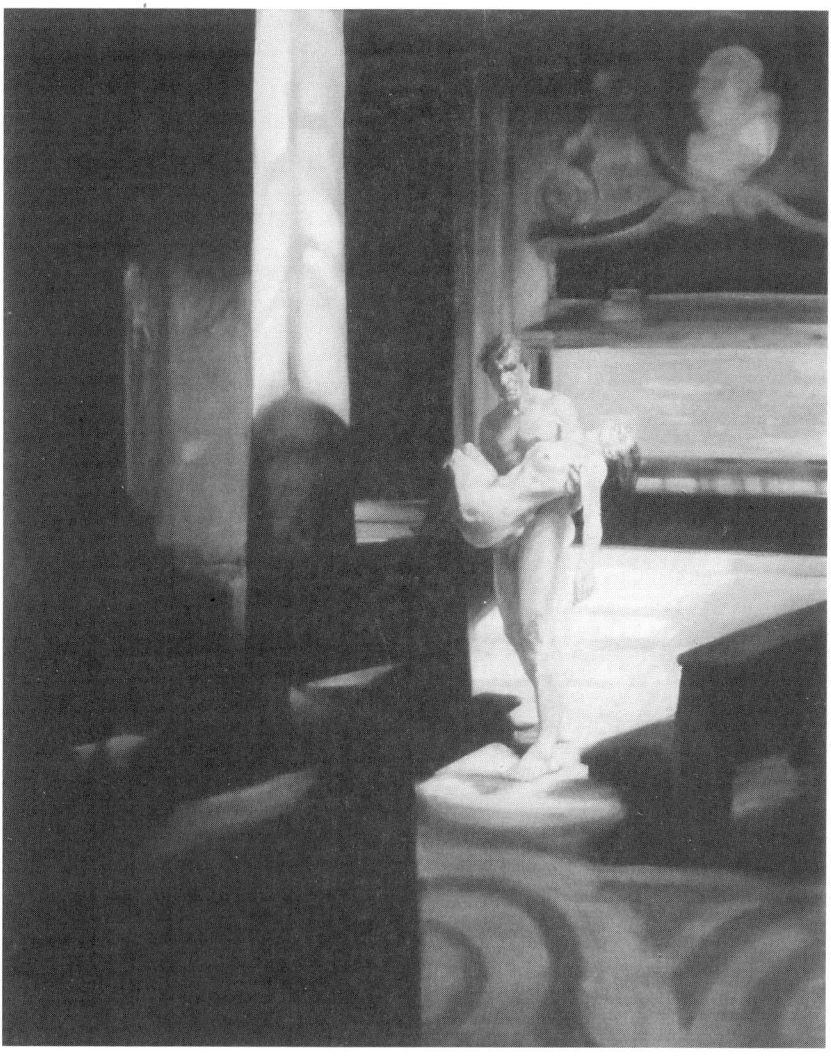

FIGURE 35.2. Eric Fischl. ***Once Where We Looked to Put Down Our Dead***, 1996. 98" × 80". Oil/linen. Collection: The Broad Art Foundation, Santa Monica, California. Used by permission of the artist. PHOTOGRAPH COURTESY OF MARY BOONE GALLERY, NEW YORK.

is and however difficult it is, is struggling with what he knows he must do. The man must summon the courage to leave his woman.

With *Once Where We Looked to Put Down Our Dead*, Fischl also shows us his own fortitude in a time when many pictures by painters focus on the shabby side of human life. Here it is not Fischl's pluck in seeing a lot in a sleazy subject, as with *Bad Boy*. Here he shows something that we have seen little of from late-twentieth-century painters: he directly portrays human life as having a tragic aspect.

*Cyclops Among the Eternally Dead*, 1995 (Figure 35–3), seems a wry commentary on what we see and what we don't see. The central figure is a middle-aged man — perhaps a churchman — dressed in black. Unseeing,

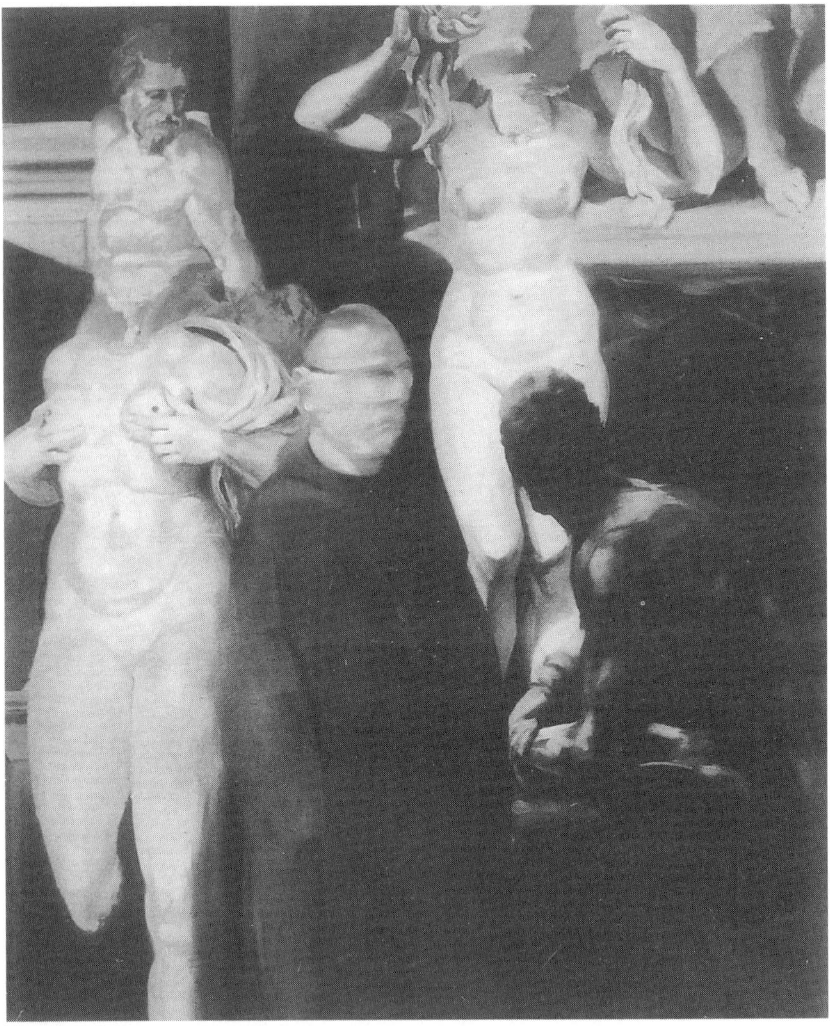

**FIGURE 35.3. Eric Fischl. *Cyclops Among the Eternally Dead*, 1996. 65" × 55". Oil/linen. Used by permission of the artist.** PHOTOGRAPH COURTESY OF MARY BOONE GALLERY, NEW YORK.

he walks past two statues of voluptuous women, one of whom stands in the gaze of a statue of a man. Like a man with no depth to his one-eyed vision, the central churchman is Cyclops. Many men, even those who are no more active sexually than a statue, are ever so keen to deep-focus their two eyes to try to expose the bodies beneath young women's clothes, even if the clothed women are solid stone.

Fischl's *Anger, Remorse, Fear or Incontinence*, 1996, shows, in a mostly dark museum, a large, brightly lit statue of a long-gone human leader or ancient god. In front of the grand statue, illuminated by a shaft of light, is a small figure, a man. The man's head is cast down, perhaps in fearful obeisance, perhaps to hide his anger. However, there happens to be a bit of a puddle on the floor beneath him. So he may be doing no more than looking at the evidence of his incontinence and fearing the embarrassing consequences. Apart from its droll humor, the title explicitly suggests what we have already seen in Fischl's uncertain-story pictures: that, although feeling of some kind comes through, it is difficult to obtain unambiguous meaning (which usually requires words) from a frozen moment in the activities of people — people in museums, people in pictures, or ordinary people living the stories of their own lives.

The artist has made some remarks about his work that suggest his own interest does not always include explicit storytelling. "Painting is a process by which I return my thoughts to feelings ... feelings that were caused by events before they become words, before they become thoughts. And because painting is a nonverbal ... condition, it's about trying to get all the words out of the picture so the picture can just represent. When someone comes to it they can start putting words back in."[6]

## *Chapter 36*

# Transmuting Commodities into Art: Jeff Koons

Take a glance or two at a couple of Jeff Koons's artworks: *New Hoover Deluxe Shampoo Polishers, New Shelton Wet/Dry 5-Gallon Quadradecker*, 1981–87 (Figure 36–1), and *Jeff in the Position of Adam*, 1990 (Figure 36–2). The first clearly shows floor-cleaning devices with fluorescent lamps, all attached to a wall-mounted piece of transparent plastic. Just as clearly, the second shows a posed erotic photograph (part of his *Made in Heaven* series) of the painter with his spouse, Cicciolina. He is bare of clothing, while she wears only some titillating garments.

What's the main difference between the two creations? The first is fatuous and vacuous. Focusing our attention on floor-cleaners only reminds us of what we well know: that too much of our limited time is spent looking at what engulfs us, advertisements for commonplace household devices. The second artwork is risqué and zesty. Although *Jeff in the Position of Adam* is only somewhat erotic, *Manet 1991*, in the same *Made in Heaven* series, can be regarded as pornographic. *Manet 1991* shows a tasteful outdoor background, in front of which the nude artist is about to press his lips between his naked wife's open legs.

Then what do the two illustrated creations have in common? (1) They are artful works, (2) they show facets of the modern life that science and technology have made possible for us, and (3) they point out (with household machines and pornographic pictures) that our world is charged with commerce and commodities. In times past such couplings would have made works like these simply unacceptable as art. "So," we viewers might well ask, "what's the deal?" Before trying to see just what the deal is, let's

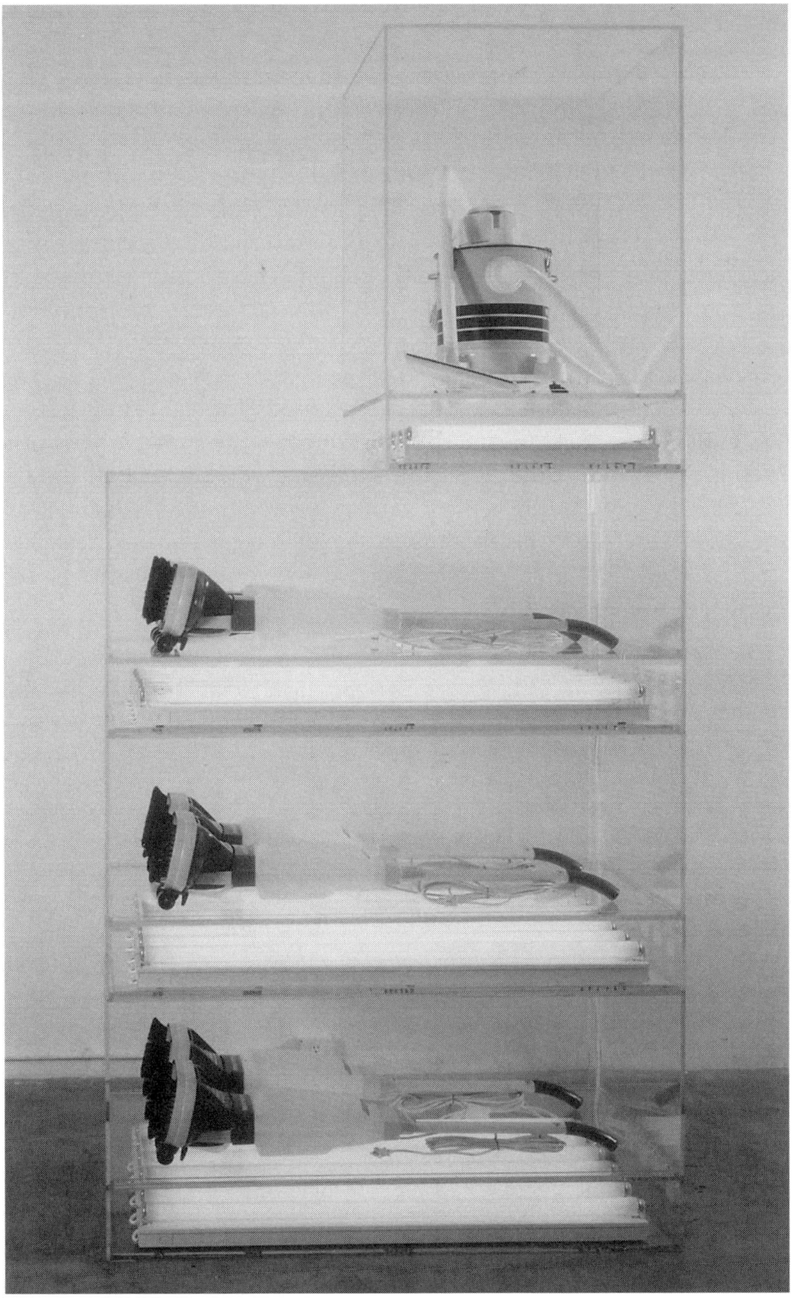

FIGURE 36.1. Jeff Koons. ***New Hoover Deluxe Shampoo Polishers, New Shelton Wet/Dry 5-Gallon Displaced Quadradecker***, 1981–87. Vacuum cleaner, six shampoo polishers, Plexiglas and fluorescent lights. 116 × 54 × 28 in. The Broad Art Foundation. 

look at some comments about Koons's work from art-world writers. Their very existence also confirms that the art world accepts his creations as real art.

Peter Schjeldahl writes: "When I first encountered a work by Koons in 1981—two vacuum cleaners encased in spotless fluorescent-lighted plexiglas—I was amazed ... Koons feeds off, and exacerbates, our culture's hysterical telescoping of all values into the seductiveness of things for sale...."[1]

From Robert Rosenblum: "In ... his most startling and publicity-hungry works, an on-going series offered under the cinematic title, 'Made in Heaven,' he and [his spouse] carried the bliss of perfect love, marriage, and sex to ... celestial climes.... [He] paraphrased the posture of Michelangelo's nude Adam, joining himself to an almost-nude Eve.... Among other amazing things about this group of images ... was that, of all unlikely things, art totally vanquished, or rather absorbed, sex."[2] "Instead of a porno show, the effect was like that of Japanese erotic prints, where the degree of stylization is so exaggerated that the sexual acrobatics as such are quickly submerged...."[3]

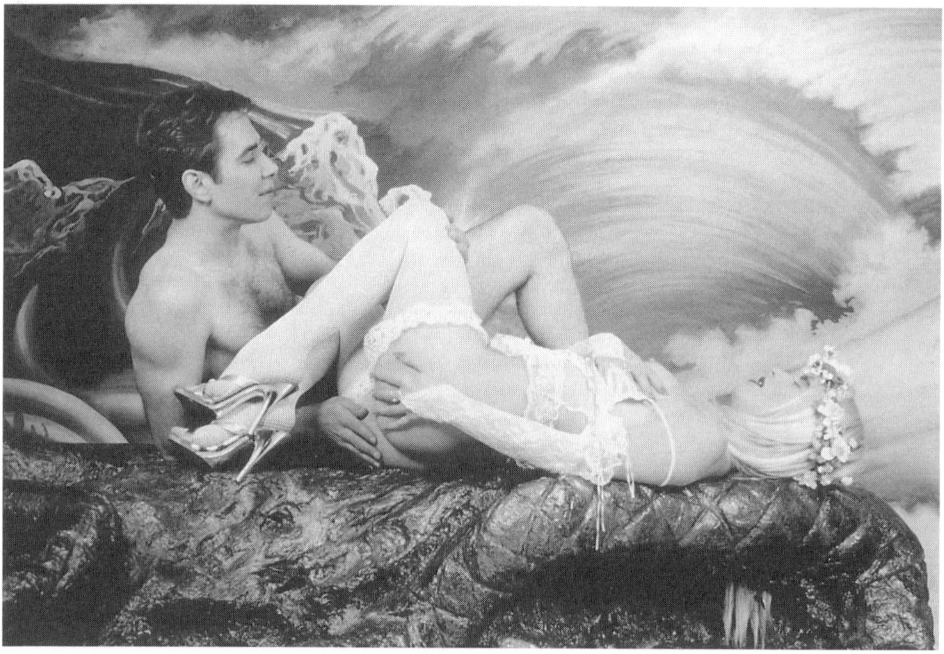

FIGURE 36.2. Jeff Koons. ***Jeff in the Position of Adam.*** **1990. Oil inks on canvas, 96 × 144 in. (243.8 × 365.8 cm.).** ©JEFF KOONS, USED BY PERMISSION. COURTESY OF GAGOSIAN GALLERY, NEW YORK.

In a book by Christopher Knight: "When it comes to sex, cleanliness isn't next to godliness in Koons's art. It *is* godliness, with all the moral anxiety inevitably attached to any conception of forbidden fruit. Koons's computer-generated paintings made from explicit photographs of sexual intercourse with his wife ... [are] plainly staged for the camera ... they have a kind of lurid cheerfulness ... 'Made in Heaven' ... is a wild oxymoron: clean pornography."[4]

For a different view of Koons's achievement, here is James Gardner. "Because Koons persists in being taken very seriously by people who take themselves more seriously still, no discussion of contemporary art can avoid him. To my knowledge he has never made a work of art that was formally or even spiritually ambitious.... Irony is the means and banality the end of Koons's art ... schlock, which deflects from itself all the sympathy for the underdog that accrues to popular art, schlock which caters to the most banal elements of middle–American complacency, seemed permanently off limits until Jeff Koons paraded it into the mainstream."[5]

With all that under our belts, let's turn to what really is a big deal—Jeff Koons's success, not as a seller of merchandise (which was his job while he worked his way into art) but as an acknowledged creator in the art world. The deal is this: Koons makes art of what is regularly pushed at us for purchase: commonplace modern artifacts and activities. His doing so is turning a nice trick, and significantly more. It is more because it is precisely the opposite of what has been happening to the world of art in recent decades: The conversion of art into the business of providing prestigious commodities that are viewed, discussed and purchased. That unspiritual reincarnation is not only what science-like objective knowledge of how to peddle has (like it or not) accomplished, but it is the source of anxiety that art itself may have been mortally wounded.

So the significance of the big Koons deal is twofold. His own work was a demonstration that art can be made from banal schlock. In return, the art community took him into its arms. It was grateful for his "not dead" verdict about art. Indeed, the art world joyfully welcomed Koons's counter-engulfing of the art part of the ever-expanding merchandising of everything known to humankind. Koons lightly trumpets his new evidence of what was and is for many a far more serious gospel—the good news that "the dead can return to life." The good news from Koons is that the subjective world of experience/feeling/soul/spirit (which after all is the basic source of art) has come back to live again as artwork! Glory, glory. Of course, ordinary advertisements (for the likes of shampoo-polishers) try—with singing praise and joyful smiles from actors playing the roles of consumers—to make the potential buyer feel good about buying. But what ads do with some success, Koons does better by selling only the *idea* of the product, and not the gul-danged thing itself. His bold apotheosis of useful spiritless devices (such as floor-cleaners) allows us to admire them with no thought of buying. This not only isolates those artifacts from their pedestrian purposes, it allows us to elevate them to the realm of art.

Turn now we should from cleaning machinery to sexual interaction. Simply stated, what most erotica and pornography does (despite its frequent sour notes of spiritless representation by actors who don't know how to act) is to provide the viewer with a means to experience, removed from its normal residence, some degree of sexual arousal. Koons does that too, and does it much better. Tunefully and with perfect pitch, he Hoovers into his bag the shameful dirt of sex in public, and then mops it out as spotless clean art.

The intimacy, in this artist's work, of commerce, sex, and art (if not otherwise apparent) is transparent in a real billboard attached to the side of a building. This "advertisement"[6] consists of (1) a huge enlargement of a picture similar to *Jeff in the Position of Adam*, and (2) the more-than-sizeable captions "MADE IN HEAVEN" and "Starring JEFF KOONS AND CICCIOLINA."

Koons's coupling of art, merchandise and pornography provides us with an occasion to note something about our basic nature. We all are often much more responsive to *representations* of things than we are to the *things* represented. On the one hand, we know quite well that we really should not "worship" images, be they religious or secular. We shouldn't allow them to be idols whose bidding commands us to be gripped by whatever they suggest. The other hand insists that, in fact, we live much more by such (subjective) portrayals than we do by the (objective) realities that are portrayed. Although we may not like it, for us to see what looks like sexual action is to be transfixed by what we see. When so clutched, we hardly care that it is simply a pair of good or bad actors who are acting out sham or real sexual interaction. The same can be said for advertising, which, as we know, is meant to command us by the sight of (what seem to be) real people showing (what are surely faked) upbeat feelings about the likes of floor-polishers.

It is instructive to compare Jeff Koons's art with that of Thomas Kinkade, a painter who makes decidedly eye-catching works. Those made by Kinkade, however, are simply not accepted in the art community as art. Both creators are firm in their efforts to ensure that there shall be no distance between art and market-artifacts. Koons, however, can be seen as an anti-matter version of Kinkade, who is an art-creator and a pusher of countless copies of his creations. Koons's thrust is to make art (which is to say, things of the "spirit") of ordinary, but up-to-date, "matter." Kinkade, however, is more than merely masterful at the conversion of heart-warming artful pictures about yesteryear's objects (such as homey thatched cottages) into modern material commodity copies. He is indeed a virtuoso at making matter out of spirit.

About Koons, an art critic writes: "Ever with his eye on the consumer, this paragon of a successful artist ... has built up year by year ... what amounts to a Jeff Koons Gift Shoppe. In his inventory of ... objects ... the range of consumer temptations is vast."[7] Kinkade, in the view of the art community, is the disbarred-from-membership Devil's Disciple. He is disallowed seemingly for not making "real" art, disrespected more likely for the gargantuan Brobdingnagian scale of his self-merchandising. Kinkade's art skills are real, but his genius is the concentrated peddling of millions of copies of his sentiment-laden paintings that show the likes of cozy little old homes near the woods.

When it comes to inventory, whatever the Koons Shoppe sells is only a few small potatoes compared with Kinkade's thousandfold-replicated sweet spuds and yummy yams. In Kinkade, we have a creator with many shops in malls and a widely broadcast TV shopping channel on which he regularly appears. By total count (some ten million) of objects sold, Kinkade dwarfs not only Koons. He makes teeny-weeny Lilliputians of all present and past painters put together.

Kinkade's art enterprise is mostly pure object-marketing, but it has an undeniable central smidgeon of cottage art (which last bit is ignored by the art world). Kinkade's copies of comfy, cozy dwellings give us pure satisfying sentimentality. They do that with nothing at all in the way of present-day citric acid to make them acceptable to the art community by kicking them up a notch before we taste their straight schmaltz. Koons's work reverses that of Kinkade. It amounts to an absorbing of modern commerce, one that slurps up and swallows down the hard lemonade of mass culture — and in the process, makes art out of merchandise that ranges from polishers to pornography. Koons offers everyone yeasty chances for plenty of peppy giggles, gurgles and hidden sneers beneath the gulps.

Jeff Koons's counter-engulfment of commerce is just the right compensation for the massive marketing that has engulfed the presentation and distribution of art itself. His acceptance by the art community can be regarded as its reaction-formation to the massive conversion of art into merchandise. The art world's response to him (using some words from Lewis Carroll) has been, "Come to my arms, my beamish boy! Thou hast slain the Jabberwock."

## *Chapter 37*

# Postmodern Pictures of Women: David Salle

After a glimpse, think about a couple of David Salle's coarse paintings: *Gericault's Arm*, 1985 (Figure 37–1), and *Blue Paper*, 1986 (Figure 37–2). The primary objects in *Gericault's Art* are two front views of an unclad young woman standing with her underpants lowered. In *Blue Paper* the viewer's attention is caught mainly by two rear views of the bare behind of a young woman bending forward. These are like many others of his, where (in the midst of unrelated objects) we find "the endless depictions of woman-models ... posed and painted ... for the spectator's gaze. More often than not they are nude or partially dressed and in sexually explicit positions—legs spread, bending over, on their knees."[1]

Among Salle's other paintings that are similar in this respect are *His Brain*, 1984, largely a rear view of one bare young woman; *Muscular Paper*, 1985, two undraped young women seen from the back; and *Rob Him of Pleasure*, 1979, which includes several views of unclad young women.

A second glance at the two illustrations tells you only a bit more. Although "the chilling banality of pornography is preserved ... [Salle] has abandoned pornography's descriptive detail of genitalia."[2] So we have to ask, just what is going on here? What kind of ideas is the painter putting into our minds? Is Salle providing some kind of a strip *tease*? After all, he shows women "stripped" in whole or in part, while he apparently holds back the explicit view of their primary external sexual parts.

Well, no, these pictures may be erotic teases, but they are not really strip teases. Helpful to us is the fact that David Salle admires irony. He values it "because of its ability to admit complication and progress ... of

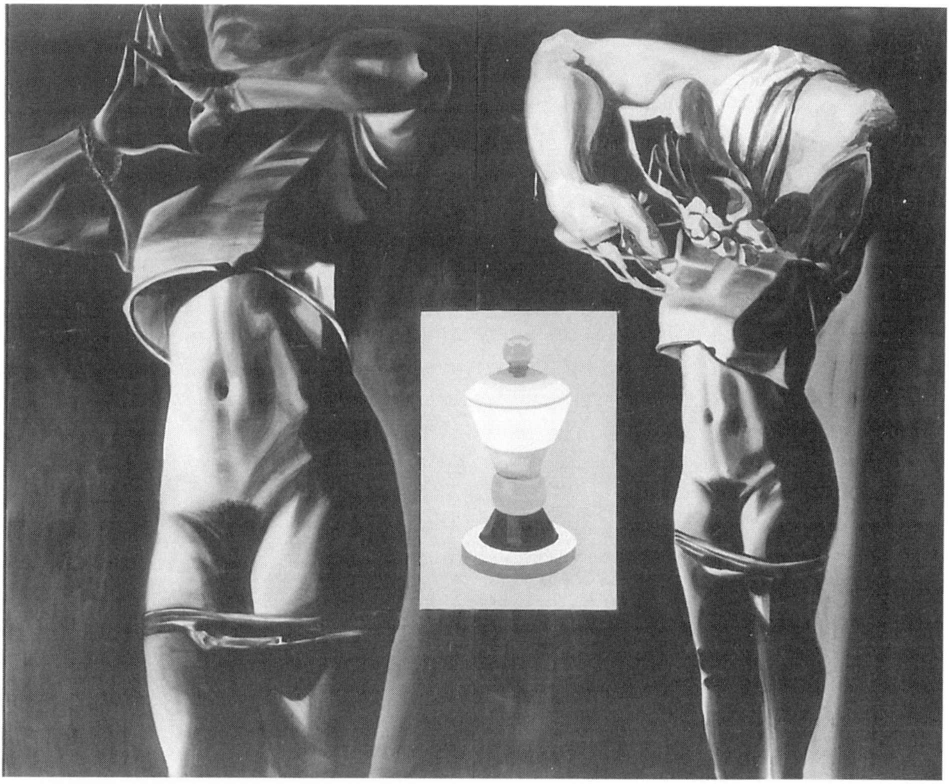

FIGURE 37.1. David Salle (b. 1952). ***Gericault's Arm***, **1985. Oil and synthetic polymer paint on canvas, 6' 5⅞" × 8' ¼". Gift of the Louis and Bessie Adler Foundation, Inc., Seymour M. Klein, President; Agnes Gund; Jerry I. Speyer Fund; and purchase. (285.1986). The Museum of Modern Art, New York, NY, U.S.A. Art ©David Salle/Licensed by VAGA, New York, NY. DIGITAL IMAGE ©THE MUSEUM OF MODERN ART/LICENSED BY SCALA/ART RESOURCE, NY.**

thinking."[3] What if this artist's images of women are *ironic* pictures? That would mean they are each a "dissimulation," which in turn means "a hiding under a false appearance." But just what is the false appearance — the women as they are shown? And what is the real thing that is hidden? Surely not the paintbrush touches that might have detailed their genitals. We will see that there is more here. And in a significant way, there is less.

Let's start our pursuit of the meaning of Salle's female images by comparing them with unclothed-woman paintings by his predecessors a century or so ago. *Origins of the Universe* is the pretentious title of an 1866 painting by Gustave Courbet. This clearly erotic work is a crotch shot, a faceless front view of the legs-apart primary private parts of a young

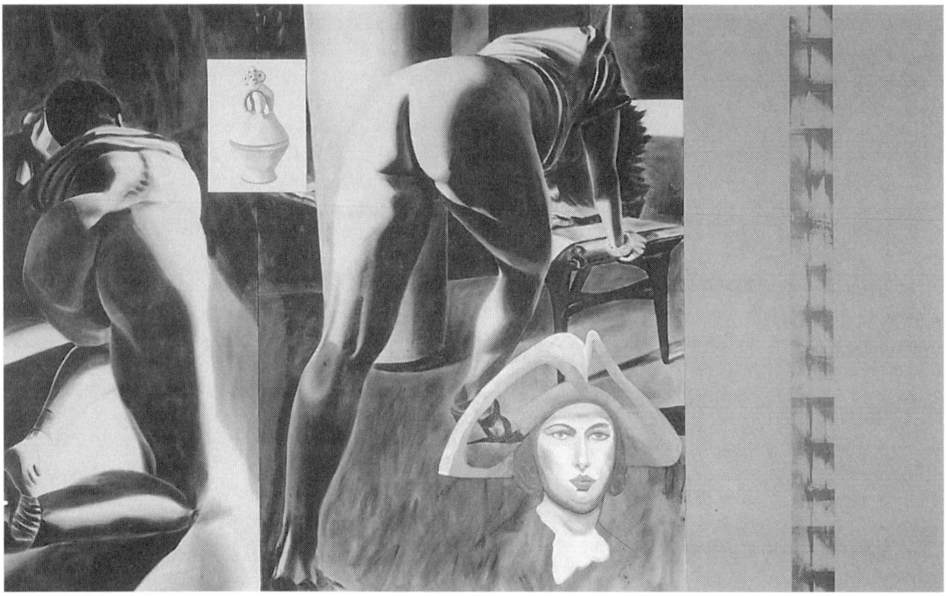

FIGURE 37.2. David Salle. *Blue Paper*, 1986. Acrylic and oil on canvas, 108 × 174 in. (274 × 442 cm.) Art ©David Salle/Licensed by VAGA, New York, NY. COURTESY GAGOSIAN GALLERY, NEW YORK.

woman. Edouard Manet's *Olympia* (1863) shows, staring back at us, a naked young provider of sexual pleasure. Although she does not provide her viewers with access to the object of Courbet's explicit gaze, Olympia is surely erotic. In 1879, Edgar Degas painted *Party at the Brothel.* It is only marginally erotic because the no-longer-young unclothed sex-workers are not nearly as attractive as they were in their prime. In 1906, Paul Cézanne gave us *Large Bathers II.* Its bare bulky women with all-alike expressionless appearance puts these images near the bottom on the scale of eroticism. Pablo Picasso's 1907 *Les Demoiselles d'Avignon* (The Young Ladies of Avignon) shows the viewer a group of unclothed in-house prostitutes. Yet their sexuality is so lacking in erotic expression that the picture would get from prurient viewers, even if they are adolescent boys, only yawns.

What is clear from these few paintings is that pictures in the past of uncovered young women cover a whole range of eroticism. If Salle's pictures are about eroticism, they should fit somewhere in this range. But they hardly straddle its middle or anywhere else. So we must conclude that they are not just about sexiness as such. What then? We can get further help — limited help — from the artist. Here's how he explains what he sees as central to his pictures. "Everything in this world is simultaneously itself and a representation of the idea of itself. This was in a sense my big art epiphany.... [You see the thing itself] but you also see the thing ... as all of the ideas about it — as a representation."[4]

What came to Salle was the *explicit* insight that we see things in the world in this double-edged way. This is like what René Magritte tried to convey some seventy-five years ago with his *The Treachery of Images*. There he applied some paint to appear like a "pipe," and more paint to shape words which (translated to English) read, "This is not a pipe." (He didn't hammer his thought by adding, "It is my idea of what a pipe looks like.") That suggests that Salle's paintings might be (1) views of young women as *themselves* (persons with their own lives), and also (2) views of young women that represent *ideas* (in this case, notions about females as erotic objects).

That some of our views about young women have these two aspects is hardly earth-shaking news. Indeed, we humans may well have acted on such non-explicit twofold knowledge even before we had words. For easy example of such implicit two-sided thinking in the world of painted young women, consider another picture Courbet made (a decade before he created his crotch shot): *Interior of My Studio....* It shows the painter at work on a canvas in his studio, where he is surrounded by friends and admirers. Behind the artist stands a young female model who is unclothed. As *herself*, she is shown looking with close interest at the landscape he is creating. For a viewer of *Interior of My Studio...*, however, a glance at her attractive body also raises the *idea* of an erotic sex object. It was Courbet, of course, who had it both ways by giving us both views.

Manet used his *Olympia* model to pose for his *Déjeuner sur l'herbe* (Luncheon on the Grass). There, the bare central female simply sits tranquilly as herself just having an ordinary picnic lunch with two (dressed) men and another young woman (in the background). This work, however, became a scandal of sorts because it shows viewers an ordinary woman as a naked sex object. This double-take, and *Olympia*, brought complaints and fame to Manet.

Has this look at two-sided pictures of unclothed young women helped us? Nope, in truth it has not helped at all. Apart from showing unclad women, Salle's paintings are not like those old paintings. The difference is this: The women in his pictures are *disconnected* from everything else that is shown in them. These pictures from this artist all show unrelated objects. The pictured young women are themselves shown as no more than unrelated objects. Like Magritte's picture title, Salle's art deals with *the treachery of images*, but, as we will see, it does so in a way that is distressing and despairing. These pictures from Salle are not at all *about* unadorned young women as *themselves.* They are conspicuous in showing us next to nothing at all about the women as thinking, feeling human beings with lives of their own.

Let's change our approach and grapple with Salle's paintings as entities, rather than as containers of somewhat erotic images of young women. *Gericault's Arm* shows the two images of bare women, the picture title's "arm," and a table lamp. *Blue Paper* is a picture with the three parts that make it a triptych. As we have seen, the left and center parts have rear views of a bent-forward young woman with no clothes beneath her waist. The left panel also has, superimposed, a picture of an ancient yellow covered jug. Superimposed at the right bottom of the middle panel is the face of a young woman wearing clothing and a hat as well as lipstick and cheek rouge. Central to the remaining third of the picture is a vertical stripe of successive frames of film.

If nothing else, the combination of things in each of these two pictures creates "the possibility of numerous readings.... Narrative is encouraged and forestalled, or, put another way, the possibility of narrative is indicated but not fulfilled. It is this perception of a double movement, toward a resolution ... as 'meaning,' yet away from it as things fall apart"[5] that we must cope with. Yet, double movement or not, there ought to be some central sense here—after all, these are not random paintbrush droplets. Surely it is time—indeed, past time—for the rabbit of at least some shred of significance to jump out of these disconnected-seeming paintings by Salle, and to jump into our minds. The painter provides some sorely needed help with the following comments—although they happen to be about a predecessor painter. "I was astonished by [his] paintings.... They were so perverse. I was very interested in works where you had trouble figuring out what the intention of the artist was—what it was that he was actually showing you, and what you had to make up...."[6] What if Salle himself chose to continue with and expand on that kind of perversity? What if he is maddeningly suggesting to us: "Hey, I make pictures. You want meaningful rabbits? Try making your own!"

So far, we have seen two clues to the meaning of these picture puzzles. First, Salle had a revelation. Like Magritte, he realized that things (such as the women and what-have-you in his paintings) are both themselves and representations of ideas. Second and unlike Magritte, there might be something perverse about Salle's works—their meaning may be hidden, private, undecipherable, perhaps even (perish the thought) deliberately absent.

Why would Salle make such inscrutable paintings? Their creation undoubtedly had significance for him. But that meaning is not of central importance for those who look at the pictures. Far more fruitful for us will be the question: Why would the art community take unto itself such strange pictures? Here's an explanation. David Salle's pictures prove to be a way of displaying the art world's anxiety and disillusion regarding (brace yourself) a big subject: changes in our world that resulted from science and its low-brow relatives. This refers to art's century-and-a-half sometimes-loving affair with science, technology and more abrasive forms of objective knowledge.

That's the relationship that started as an exciting liaison — Impressionism — in the middle of the nineteenth century. That was when the art world and other parts of the humanities were attracted to all the stimulating change that grew from science and technology. In time, that became the two cultures' established light-on-love habit of living more or less together, the custom that deteriorated in the twentieth century into a marriage of inconvenience. Although the art world was obliged to live with the other one, it soon spurned intimacy — see Expressionism's turn inward and away from the outer world where science lived.

The roundabout road to making sense of Salle's work leads us to what is called *Postmodernism*, which started in the latter part of the twentieth century. In large part, it was the result of "dismay at the side-effects of modern technology.... The resultant fragmentation of the ... Modernist dream precipitated a pessimistic retreat into the view that there was no such thing as certain knowledge."[7] As a result, the two cultures, the humanities (which includes the various arts) and science, "were no longer on speaking terms.... Enlightenment thinkers believed we can know everything ... radical postmodernists believe we can know nothing."[8] "The Enlightenment ... [was] intent upon building a sound body of knowledge about the world the human race confronts ... Postmodernism defines itself, in large measure, as [suggesting that] such a project is inherently futile, self-deceptive...."[9] "Postmodernism ... was [also] used to imply that the emphasis on originality and authorship ... is now *passé*: the work of art only acquires ... meaning according to the context within which it is placed.... Appropriation of existing images is entirely permissible...."[10]

Why on earth did the art community come to an involvement in those strange and sad postmodern judgments about our knowledge of the world we live in? The matter-of-fact reason is both plain and unpleasant. The low-brow aspects of the *other* culture — the one that centers on science — came to dominate the world of art. Put baldly, art became a big business that used semi-scientific marketing methods to sell — perhaps glorious visions of the human soul, but surely — the viewing of paintings and the artifacts themselves. Displaying art and selling it took objective knowledge of what works in the marketplace and what doesn't. One example: Peddling Vincent van Gogh (who sold only one picture while he lived) generated over a hundred million dollars of revenue with a four-month run of his pictures in a big-city museum near the end of the twentieth century.

It is true enough that we in the West have all become consumers, sometimes overstuffed ones, of everything in the way of objects. But our condition (part-time consumers) is hardly as exasperating as the circumstance of those who live in the art world. If you live there, you find that what is central to your life — art itself — has often become mere merchandise. To earn your daily bread, there are days when all you do is butter up painted canvas and funnel it down consuming throats.

Stuck in the grip of a loveless marriage, the art world welcomed pictorial protestations that life had become *meaningless*. That is the reason it was enticed and excited by Salle's art, which was postmodern and ironic in that sour sense.

With all that under our belts, we can think again about David Salle's paintings. The artist certainly has success with his pictured nakedness generating *ideas*. His paintings do indeed bring positive, negative or simply stale concepts about bare young women into our minds. For example, these scenes of unclothed young women might be an ironic view of how common ideas about sex objects have expanded to produce the big business of displaying female bodies in pornography and elsewhere. What, however, does Salle show us about the young women as *themselves*? By showing women only in the context of disconnected things, his pictures are perversely postmodern. To expropriate Gertrude Stein's words, *there is simply no there there* about women as human beings engaged with the modern world.

It is true enough that the promised land of the Enlightenment — the over-promised land where science would deliver everything as wonderfully knowable and helpful — was a dream never fulfilled. Postmodernism, however, over-reacted by going overboard the other way. It suggests this: As no knowledge is certain, why bother much with any of it? Why waste effort to make visible, for obvious example, how women have changed from those who posed for Courbet and Manet? Why try to depict modeled women interacting in exciting new ways with their world? The resentful and sterile aspect of Postmodernism casts doubt on whether all that is achievable.

Apart from openly showing women as semi-erotic objects, Salle's pictures are only indirectly about women. What is hidden is the perverse postmodern implication that knowledge (by men, to be sure) about women — women as themselves real sexual beings, let alone as real human beings — is not worth seeking.

## *Conclusions*

# Images of Our Anxieties

The Bible cautions us not to make graven images or likenesses of things in heaven or on the earth. That warning came long before art got modern, back in days of old when much of art portrayed Heaven's dwellers or their earthly representatives. So one of the reasons for that wariness about images may be this: Getting absorbed in the pictures as such makes it too easy to overlook what they represent. With modern art, the problem with depiction is a different one. Although it remains easy for us to get engrossed in the images, it is now often difficult to know just what on earth or in Heaven is beneath surface appearances.

This book has tried to show that, beneath its surface, modern art has pictured our conflicts about the results of science and technology. Consider first an ancient site that paleo-anthropologists have uncovered in Kenya, one that contains some of such results from long before we had the Bible. At that place some ten thousand chiseled stone hand-axes were found within a small area.[1] The evidence shows they were made by members of our predecessor species, *Homo erectus*. Beyond pointing to the learned skill of fabricating some of our earliest artifacts—hand-axes—such a find tells us two more things. The discovery (of ten thousand axes found together) implies that those early people had learned to specialize in their labor. It also suggests that axes were traded as commodities, exchanged perhaps for food or for animal skins. Tool-making, specialization and trading were the result of useful knowledge shared by individuals, thus of science-like objective knowledge gained in ancient days.

In the last century or two, such knowledge and its products have not only grown but have transformed our human world. One positive result has been the vast increase in the duration and comfort of life for us. Long ago, "old age" was a concept that was empty of embodying examples— disaster or disease killed people before they got old. Now, poor people in the West receive better medications than rich ones did a century ago. In our wealthy Western world, we live with an abundance, a relative-to-elsewhere glut, that we enjoy — and we also scorn. As we know, there have likewise been less desirable consequences of modern science's ever-expanding makeover of the world. Artifacts have become so plentiful that we have become alarmed that material and materialism are overwhelming us. We have rightly learned to fear the expanding power we have developed to do ourselves great harm. Many of us have also acquired a more personal anxiety — a dread that we ourselves may be no more than mortal objects not unlike those that science studies.

Modern art pictured much of that. In doing so, it also increasingly divided its attention between looking at the world outside, and turning inward and away from that world. As times changed, Manet was among the first to emulate science's search for truth and openness to innovation. Visible brushwork was truthful, likewise the truth was that men did look at young women as sex objects. The haste required by Monet and Renoir — for the new method of outdoor painting of impressions— served to represent the very idea of the rapid change of their world. Morisot reflected the transforming world by picturing some fragile moments just before they disappeared. Degas took time inside his studio to show another way in which attitudes toward women were changing — women were seen not only as sex objects but as equally human with men. Van Gogh and Gauguin turned further toward inner experience and away from troublesome outer reality, one with moving color and the other with primitive romance. Cézanne worked on the opposite side of the street; in some of his work his timeless objectivity pictured the indifferent external world. Matisse turned more to the inside by emphasizing the decorative and diminishing the realistic. Picasso and Duchamp swiveled back to the outside, one with science-emulating inventive Cubism and the other with new focus on three-dimensional manufactured objects.

The strain on us, and thus in art, increased dramatically with the harsh flogging reality of the twentieth century. Hopper showed the loneliness of life in crowded new cities. Bacon painted the bleak failure of the human spirit to cope with massive inhuman behavior. Turning away from the real world, Rothko, de Kooning, Krasner, and Pollock withdrew into imagined inside worlds of colorful rectangles, subjective shapes, and tangled threads of pigment. Johns, Rauschenberg, Lichtenstein, and Warhol stepped back outside to show our world as if it were little more than shallow surface full of prosaic objects. After the horrendous mid-century Cold

War came to an unexpected peaceful end, painting the new territories of inner and outer art continued in the work of Schnabel and Hirst. However, the creations of Richter, Kiefer, Fischl, Koons and Salle show art's halves coming together once more to mirror fuller views of the tensions of modern human life.

Although life has always required us to cope with changing times, recent rapid metamorphosis transformed the very scale of such change. It was hard to brave the new world overflowing with the artifacts and dangers that keep coming from science and technology. Nor was it easy to deal with fresh reasons to fear that we ourselves may be no more than mere material.

All of this suggests that art has reflected — with more truth than beauty — the scene of our modern human condition. To borrow (from American sculptor Leonard Baskin) some dramatic words, "our human frame, our gutted mansion, our enveloping sack of beef and ash is yet a glory. [Art holds] the cracked mirror up to man."[2]

# Chapter Notes

## Chapter 1

1. Kenneth Clark. *Moments of Vision and Other Essays* (New York: Harper & Row, 1981), p. 19.
2. Jacques Barzun, *The Use and Abuse of Art* (Princeton, N.J.: Princeton University Press, 1974), p. 33.
3. David Knight, *The Age of Science: The Scientific World-view in the Nineteenth Century* (New York: Basil Blackwell Inc., 1986), p. 2.
4. Brian L. Silver, *The Ascent of Science* (New York: Oxford University Press, 1998), p. 25.
5. Knight, pp. 32–36.
6. Silver, pp. 53–57.

## Chapter 2

1. Susanne K. Langer. *Mind: An Essay on Human Feeling, Vol. 1.* Baltimore, Md.: Johns Hopkins University Press, 1967, p. 84.
2. Langer, p. 115.
3. Langer, p. 124.
4. Jacques Barzun. *The Use and Abuse of Art.* Princeton, N.J.: Princeton University Press, 1974, p. 18.
5. Liz Dawtry, Toby Jackson, Mary Masterton, Pam Meecham & Paul Wood, editors. *Investigating Modern Art.* New Haven, Conn.: Yale University Press, 1996, p. 28.
6. Barzun, p. 34.

## Chapter 3

1. Roger Shattuck. *Forbidden Knowledge: From Prometheus to Pornography.* New York: St. Martin's Press, 1996, p. 332.
2. Robin Dunbar. *The Trouble with Science.* Cambridge, Mass.: Harvard University Press, 1995, p. 58.
3. Dunbar, p. 73.
4. Ernest Gellner. *Anthropology and Politics: Revolutions in the Sacred Grove.* Cambridge, Mass.: Blackwell Publishers Inc., 1995, pp. 251–52.
5. A.N. Wilson. *God's Funeral.* New York: W.W. Norton, 1999, p. 313.
6. Richard Maurice Bucke, M.D. *Cosmic Consciousness.* New York: E.P. Dutton, 1969, p. 9.
7. Bucke, p. 17.
8. Vern L. Bullough & Bonnie Bullough. *Sexual Attitudes: Myths & Realities.* Amherst, N.Y.: Prometheus Books, 1995, p. 13.
9. Edward O. Wilson. *Consilience: The Unity of Knowledge.* New York: Vintage Books, 1999, p. 66.
10. Gerald M. Edelman and Giulio Tononi. *A Universe of Consciousness: How Matter Becomes Imagination.* New York: Basic Books, 2000.

## Chapter 4

1. David Knight. *The Age of Science: The Scientific World-view in the Nineteenth Century.* New York: Basil Blackwell Inc., 1986, p. 5.

2. Herbert J. Muller. *The Children of Frankenstein*. Bloomington: Indiana University Press, 1970, p. 38.

3. Robin Dunbar. *The Trouble with Science*. Cambridge, Mass.: Harvard University Press, 1995, pp. 171–72.

## Chapter 5

1. Abigail Solomon-Godeau. *Male Trouble: A Crisis in Representation*. New York: Thames & Hudson, 1997, p. 61.

2. Michel Foucault. *The History of Sexuality*. New York: Pantheon Books, 1985, p. 215.

3. Eric Flaum, with David Pandy. *The Encyclopedia of Mythology*. Philadelphia: Running Press Book Publishers, 1993, p. 150.

4. John Rewald. *The History of Impressionism*. New York: The Museum of Modern Art, 1973, p. 21.

5. Frank Anderson Trapp. *The Attainment of Delacroix*. Baltimore, Md.: The Johns Hopkins University Press, 1970, p. 78.

6. Trapp, p. 322.

7. Barthelemy Jobert. *Delacroix*. Princeton, N.J.: Princeton University Press, 1997, p. 92.

8. Jobert, p. 9.

9. Trapp, p. 98.

10. Rewald, p. 17.

11. William Ian Miller. *The Anatomy of Disgust*. Cambridge, Mass.: Harvard University Press, 1997, p. 21.

## Chapter 6

1. Oscar Handlin. "Science and Technology in Popular Culture," in *Science and Culture*, edited by Gerald Holton. Cambridge, Mass.: Houghton Mifflin Company, 1965, p. 190.

2. Anne Taylor Hanson. *Manet and the Modern Tradition*. New Haven, Conn.: Yale University Press, 1977, p. 15.

3. Werner Haftmann. *Painting in the Twentieth Century*. New York: Frederick A. Praeger, 1960, pp. 26–28.

4. Haftmann, p. 49.

5. Francis Frascina, Nigel Blake, Briony Fer, Tamar Garb & Charles Harrison. *Modernity and Modernism: French Painting in the Nineteenth Century*. New Haven, Conn.: Yale University Press, 1993, p. 56.

6. Frascina et al., p. 80.

7. James H. Rubin. *Impressionism*. London: Phaidon Press Limited, 1999, p. 125.

8. Norma Broude. *Impressionism: A Feminist Reading*. New York: Rizzoli International Publications, Inc., 1991, p. 8.

9. Rubin, p. 317.

## Chapter 7

1. Francis Frascina, Nigel Blake, Briony Fer, Tamar Garb & Charles Harrison. *Modernity and Modernism: French Painting in the Nineteenth Century*. New Haven, Conn.: Yale University Press, 1993, p. 84.

2. Frascina et al., p. 85–86.

3. Peter Gay. *Pleasure Wars*. New York: W.W. Norton & Company, 1998, p. 8.

4. Frascina et al., p. 97.

5. Frascina et al., pp. 99-101.

6. Gaeton Picon. *The Birth of Modern Painting*. New York: Rizzoli International Publications, Inc., 1978, p. 19.

7. Erika Langmuir and Norbert Lynton. *The Yale Dictionary of Art and Artists*. New Haven, Conn.: Yale University Press, 2000, p. 430.

8. Kenneth Clark. *The Nude*. New York: Pantheon Books, 1956, p. 164.

9. Bram Dijkstra. *Idols of Perversity: Fantasies of Feminine Evil in Fin-de-Siècle Culture*. New York: Oxford University Press, 1986, p. 355.

10. Frascina et al., p. 279–80.

11. Dijkstra, p. 211.

12. Frascina et al., p. 21.

13. Picon, p. 24.

14. Michael Crichton. *Jasper Johns*. New York: Harry N. Abrams, Inc., 1977, p. 83.

15. Peter Schjeldahl. "The Urbane Innocent." *The New Yorker*, Nov. 20, 2000, p. 104.

## Chapter 8

1. Herbert Read. *A Concise History of Modern Painting*. New York: Thames & Hudson, Inc., 1998, p. 16.

2. Francis Frascina, Nigel Blake, Briony Fer, Tamar Garb & Charles Harrison. *Modernity and Modernism: French Painting in the Nineteenth Century*. New Haven, Conn.: Yale University Press, 1993, pp. 167–68.

3. Lawrence Hanson. *Renoir: The Man, the Painter, and His World*. New York: Dodd, Mead & Co., 1968, pp. 24–26.

4. James H. Rubin. *Impressionism*. London: Phaidon Press Limited, 1999, p. 103.
5. Meyer Schapiro. *Impressionism: Reflections and Perceptions*. New York: George Braziller, Inc., 1997, p. 51.
6. Frascina et al., p. 168.
7. Erika Langmuir & Norbert Lynton. *The Yale Dictionary of Art and Artists*. New Haven, Conn.: Yale University Press, 2000, p. 345.

## Chapter 9

1. Julian Barbour. *The End of Time: The Next Revolution in Physics*. New York: Oxford University Press, 2000, p. 18.
2. Jean Dominique Rey. *Berthe Morisot*. New York: Crown Publishers, Inc. (undated), p. 50.
3. Philippe Huisman. *Morisot: Enchantment*. New York: French & European Publications, Inc., 1963, p. 20.
4. Rey, p. 21.
5. Bradd Shore. *Culture in Mind: Cognition, Culture, and the Problem of Meaning*. New York: Oxford University Press, 1996, p. 15.
6. Barbara Ehrlich White. *Impressionists Side by Side*. New York: Alfred A. Knopf, 1996, p. 216.
7. White, p. 219.

## Chapter 10

1. Ian Dunlop. *Degas*. New York: Galley Press, 1979, p. 125.
2. Philippe Huisman. *Morisot: Enchantment*. New York: French & European Publications, 1963, p. 44.
3. Bram Dijkstra. *Idols of Perversity: Fantasies of Feminine Evil in Fin-de-Siècle Culture*. New York: Oxford University Press, 1986, p. 181.
4. Daniel Catton Rich. *Degas*. New York: Harry N. Abrams, Inc., 1952, p. 5.
5. Eunice Lipton. *Looking into Degas*. Berkeley: University of California Press, 1986, p. 88.
6. Dunlop, p. 134.
7. James H. Ruben. *Impressionism*. London: Phaidon Press Limited, 1999, p. 203.
8. Lipton, p. 134.
9. Lipton, p. 165.
10. Rich, p. 6.
11. Dijkstra, p. 119.
12. Dunlop, p. 144.
13. Denys Sutton. *Edgar Degas, Life and Work*. New York: Rizzoli International Publications, Inc., 1986, p. 251.

## Chapter 11

1. René Huyghe. *Van Gogh*. New York: Crown Publishers, Inc., 1988, p. 48.
2. Huyghe, p. 48.
3. Meyer Schapiro. *Impressionism*. New York: George Braziller, 1997, p. 316.
4. Isaac Asimov. *The Universe*. New York: Walker and Company, 1966, p. 3.
5. Robert Jastrow. *God and the Astronomers*. New York: W.W. Norton & Company, 1978, pp, 113-15.
6. Huyghe, p. 5.
7. Joseph-Emile Muller. *Fauvism*. New York: Frederick A. Praeger, 1967, p. 19.
8. Bernard Denvir. *Post-Impressionism*. New York: Thames & Hudson, Inc., 1992, p. 175.
9. Huyghe, p. 15.
10. Huyghe, p. 65.
11. Huyghe, pp. 83, 86.
12. Norbert Lynton. "Expressionism," in *Concepts of Modern Art*, edited by Nikos Stangos. New York: Thames & Hudson, Inc., 1994, p. 32.

## Chapter 12

1. Belinda Thomson. *Gauguin*. New York: Thames & Hudson, Inc., 1987, p. 114.
2. Stephen F. Eisenman. *Gauguin's Skirt*. New York: Thames & Hudson, Inc., 1997, p. 139.
3. Eisenman, p. 33.
4. Charles Harrison, Francis Frascina & Gill Perry. *Primitivism, Cubism, Abstraction: The Early Twentieth Century*. New Haven, Conn.: Yale University Press, 1993, p. 30.
5. Eisenman, p. 34.
6. Eisenman, p. 56.
7. Nancy Mowll Mathews. *Paul Gauguin: An Erotic Life*. New Haven, Conn.: Yale University Press, 2001, p. 62.
8. Eisenman, p. 124.
9. Eisenman, p. 130.
10. Basil Taylor. *The Impressionists and Their World*. London: Phoenix House, 1953, p. 15.
11. Eisenman, p. 15.

12. Thomson, p. 199.
13. Thomson, p. 194.
14. Eisenman, p. 135.
15. Eisenman, pp. 143–44.
16. Eisenman, p. 147.
17. Eisenman, p. 92.

## Chapter 13

1. Herbert Read. *A Concise History of Modern Painting*. New York: Thames & Hudson, Inc., 1998, p. 13.
2. Read, pp. 20–21.
3. Ivan Taillandier. *Cézanne*. New York: Crown Trade Paperbacks, 1977, pp. 86–87.
4. Colin Wilson. *Alien Dawn*. New York: Fromm International Publishing Corp., 1988, p. 296.
5. Robert E. Bartholomew & George S. Howard. *UFOs & Alien Contact: Two Centuries of Mystery*. Amherst, N.Y.: Prometheus Books, 1998, p. 21.
6. Richard W. Murphy and the Editors of Time-Life Books. *The World of Cézanne: 1839–1906*. New York: Time-Life Books, 1972, p. 114.
7. Murphy et al., p. 102.
8. Murphy et al., p. 10.
9. Bradd Shore. *Culture in Mind: Cognition, Culture, and the Problem of Meaning*. New York: Oxford University Press, 1996, p. 15.
10. Robert Hughes. *The Shock of the New*. New York: Alfred A. Knopf, 1996, pp. 125–26.
11. Taillandier, pp. 6–8.
12. Barbara Ehrlich White. *Impressionists Side by Side*. New York: Alfred A. Knopf, 1996, p. 107.
13. Murphy, p. 9.
14. Taillandier, p. 5.
15. Taillandier, p. 5.
16. James H. Rubin. *Impressionism*. London: Phaidon Press Limited, 1999, p. 387.

## Chapter 14

1. Roger Shattuck. *Forbidden Knowledge: From Prometheus to Pornography*. New York: St. Martin's Press, 1996, p. 302.
2. Brian L. Silver. *The Ascent of Science*. New York: Oxford University Press, 1998, p. 130.
3. Eric W. Gritsch. *Bornagainism: Perspectives on a Movement*. Philadelphia: Fortress Press, 1982, p. 29.
4. Gritsch, p. 41.
5. Stephen Jay Gould. *Rocks of Ages: Science and Religion in the Fullness of Life*. New York: Random House, 2002.
6. Herbert Read. *A Concise History of Modern Painting*. New York: Thames & Hudson, Inc., 1985, p. 105.
7. John Elderfield. *The "Wild Beasts": Fauvism and Its Affinities*. New York: Oxford University Press, 1976, p. 97.

## Chapter 15

1. Robert Hughes. *The Shock of the New*. New York: Alfred A. Knopf, 1996, p. 134.
2. Carol Strickland & John Boswell. *The Annotated Mona Lisa*. Kansas City, Mo.: Andrews McMeel Publishing, 1992, p. 134.
3. Herbert Read. *A Concise History of Modern Painting*. New York: Thames & Hudson, Inc., 1985, p. 38.
4. Strickland & Boswell, p. 134.
5. Werner Haftmann. *Painting in the Twentieth Century*. New York: Frederick A. Praeger, 1960, p. 52.
6. Leo Steinberg. *Other Criteria: Confrontations with Twentieth Century Art*. New York: Oxford University Press, 1972, p. 3.
7. Steinberg, p. 7.
8. Susanne K. Langer. *Mind: An Essay on Human Feeling, Vol. 1*. Baltimore, Md.: Johns Hopkins University Press, 1967, p. 312.
9. Charles Harrison, Francis Frascina & Gill Perry. *Primitivism, Cubism, Abstraction: The Early Twentieth Century*. New Haven, Conn.: Yale University Press, 1993, p. 53.
10. Kenneth Clark. *The Nude*. New York: Pantheon Books, 1956, p. 8.
11. Harrison et al., p. 58.
12. Hermann Leight. *History of the World's Art*. London: George Allen and Unwin, Ltd., 1952, p. 304.
13. Jack Flam. *Matisse on Art*. Berkeley: University of California Press, 1995, p. 32.

## Chapter 16

1. Nikos Stangos, editor. *Concepts of Modern Art: From Fauvism to Postmodernism*. New York: Thames & Hudson, Inc., 1994, p. 9.
2. John Golding. *Cubism*. Cambridge, Mass.: Harvard University Press, 1988, p. 70.
3. Hermann Leight. *History of the*

*World's Art*. London: George Allen and Unwin, Ltd., 1952, p. 307.

4. Werner Haftmann. *Painting in the Twentieth Century*. New York: Frederick A. Praeger, 1960, p. 109.

5. Erika Langmuir & Norbert Lynton. *The Yale Dictionary of Art and Artists*. New Haven, Conn.: Yale University Press, 2000, p. 531.

6. William Ian Miller. *The Anatomy of Disgust*. Cambridge, Mass.: Harvard University Press, 1997, p. 54.

7. Liz Dawtrey, Toby Jackson, Mary Masterton, Pam Meecham & Paul Wood, editors. *Investigating Modern Art*. New Haven, Conn.: Yale University Press, 1996, p. 65.

8. Bram Dijkstra. *Idols of Perversity: Fantasies of Feminine Evil in Fin-de-Siècle Culture*. New York: Oxford University Press, 1986, p. 320.

9. Robert Rosenblum. *Cubism and Twentieth-Century Art*. New York: Harry N. Abrams, Inc., 1966, p. 72.

## Chapter 17

1. Matthew Collings. *This Is Modern Art*. New York: Watson-Guptill Publications, 2000, p. 198.

2. Arthur C. Danto. *The Philosophical Disenfranchisement of Art*. New York: Columbia University Press, 1996, p. 63.

3. Ibid., p. 32.

4. Arthur C. Danto. *The Transfiguration of the Commonplace*. Cambridge, Mass.: Harvard University Press, 1981, p. 5.

5. Ibid., p. 63.

6. Edward Lucie-Smith. *Visual Arts in the Twentieth Century*. New York: Harry N. Abrams, Inc., 1997, p. 101.

7. Susanne K. Langer. *Mind: An Essay on Human Feeling, Vol. 1*. Baltimore, Md.: Johns Hopkins University Press, 1967, p. 128.

8. *The Jerusalem Bible*. New York: Doubleday, 1996, p. 102.

9. Erika Langmuir & Norbert Lynton. *The Yale Dictionary of Art and Artists*. New Haven, Conn.: Yale University Press, 2000, p. 210.

10. Clifford Geertz. *The Interpretation of Cultures*. New York: Basic Books, 1973, p. 434.

## Chapter 18

1. Herbert J. Muller. *The Children of Frankenstein*. Bloomington: Indiana University Press, 1970, p. 338.

2. Eric W. Gritsch. *Born Againism: Perspectives on a Movement*. Philadelphia: Fortress Press, 1982, pp. 40–46.

3. Gritsch, p. 49.

4. Brian L. Silver. *The Ascent of Science*. New York: Oxford University Press, 1998, p. 492.

5. Silver, p. 493.

6. Edward Lucie-Smith. *Movements in Art Since 1945*. New York: Thames & Hudson, Inc., 1995, p. 35.

7. Norbert Lynton. "Expressionism," in *Concepts of Modern Art*, edited by Nikos Stangos. New York: Thames & Hudson, Inc., 1994, pp. 39–43.

8. Werner Haftmann. *Painting in the Twentieth Century*. New York: Frederick A. Praeger, 1960, p. 21.

## Chapter 19

1. James H. Rubin. *Impressionism*. London: Phaidon Press Limited, 1999, p. 81.

2. Robert Hobbs. *Edward Hopper*. New York: Harry N. Abrams, Inc., 1987, p. 139.

3. Arthur C. Danto. *After the End of Art*. Princeton, N.J.: Princeton University Press, 1997, p. 118.

## Chapter 20

1. Paul Vogt. *Expressionism: German Painting, 1905–1920*. New York: Harry N. Abrams, Inc., 1980, p. 7. The quote is from Hermann Bahr's description of the period from 1905 to 1920.

2. M. H. Abrams, gen. editor. *The Norton Anthology of English Literature*, Vol. 2. New York: W.W. Norton & Co., 1974, p. 1925.

3. Carol Strickland & John Boswell. *The Annotated Mona Lisa*. Kansas City, Mo.: Andrews McMeel Publishing, 1992, p. 163.

4. David Sylvester. *Looking Back at Francis Bacon*. New York: Thames & Hudson, Inc., 2000, pp. 18–21.

5. Sylvester, p. 49.

6. Sylvester, p. 29.

## Chapter 21

1. Christopher Cerf & Victor Navasky. *The Experts Speak*. New York: Villard Books, 1998, p. 307.
2. Christopher Knight. *Last Chance for Eden: Selected Art Criticism, 1979–1994*. Edited by Malin Wilson. Los Angeles: Art Issues Press, 1995, p. 178.
3. Knight, p. 240.
4. Knight, p. 240.
5. Knight, p. 242.
6. Marvin Harris. *Our Kind*. New York: Harper & Row, 1989, p. 265.
7. Robert Wright. *The Moral Animal*. New York: Pantheon Books, 1994, p. 243.

## Chapter 22

1. Edward Lucie-Smith. *Visual Arts in the Twentieth Century*. New York: Harry N. Abrams, Inc., 1997, p. 227.
2. Leo Steinberg. *Other Criteria: Confrontations with Twentieth Century Art*. New York: Oxford University Press, 1972, p. 259.
3. Werner Haftmann. *Painting in the Twentieth Century*. New York: Frederick A. Praeger, 1960, p. 449.
4. Ellen R. Johnson, ed. *American Artists on Art: From 1940 to 1980*. New York: Harper & Row, 1982, p. 21.
5. Liz Dawtrey, Toby Jackson, Mary Masterton, Pam Meecham & Paul Wood, eds. *Investigating Modern Art*. New Haven, Conn.: Yale University Press, 1996, p. 66.
6. Bram Dijkstra. *Idols of Perversity: Fantasies of Feminine Evil in Fin-de-Siècle Culture*. New York: Oxford University Press, 1986, p. 249.
7. Judy Chicago and Edward Lucie-Smith. *Women and Art: Contested Territory*. New York: Watson-Guptill Publications, 1999, p. 127.

## Chapter 23

1. Ann Middleton Wagner. *Three Artists*. Berkeley: University of California Press, 1996, p. 142.
2. Edward Lucie-Smith. *Movements in Art Since 1945*. New York: Thames & Hudson, Inc., 1995, pp. 183–85.
3. Wagner, p. 155.
4. Edward Albee. "Considering Krasner." In *Lee Krasner*. Los Angeles: Tasende Gallery, 1998, p. 7.
5. Robert Nadeau & Menas Kafatos. *The Non-Local Universe: The New Physics and Matters of the Mind*. New York: Oxford University Press, 1999.
6. Nadeau & Kafatos, p. 94.
7. A.G. Cairns-Smith. *Evolving the Mind: On the Nature of Matter and the Origin of Consciousness*. New York: Cambridge University Press, 1996, pp. 264–65.
8. Nadeau & Kafatos, p. 182.
9. Wagner, p. 178.
10. Albee, p. 5.
11. Wagner, p. 121.

## Chapter 24

1. Kirk Varnedoe with Pepe Karmel. *Jackson Pollock*. New York: The Museum of Modern Art, 1998, p. 31.
2. Varnedoe, p. 59.
3. Edward Lucie-Smith. *Movements in Art Since 1945*. New York: Thames & Hudson, Inc., 1998, p. 33.
4. Werner Haftmann. *Painting in the Twentieth Century*. New York: Frederick A. Praeger, 1960, p. 448.
5. Leo Steinberg. *Other Criteria: Confrontations with Twentieth Century Art*. New York: Oxford University Press, 1972, p. 263.
6. Liz Dawtrey, Toby Jackson, Mary Masterton, Pam Meecham & Paul Wood, eds. *Investigating Modern Art*. New Haven, Conn.: Yale University Press, 1996, p. 110.
7. Dawtrey, p. 124.
8. Varnedoe, p. 47.
9. Varnedoe, p. 48.
10. Varnedoe, p. 66.

## Chapter 25

1. Brian L. Silver. *The Ascent of Science*. New York: Oxford University Press, 1998, p. 479.
2. B. Alan Wallace. *The Taboo of Subjectivity: Toward a New Science of Consciousness*. New York: Oxford University Press, 2000, pp. 27–28.
3. Wallace, p. 37.
4. Wallace, p. 88.
5. Richard Lewontin. "Darwin, Mendel, and the Mind," in *It Ain't Necessarily So: The Dream of the Human Genome and Other Illusions*. New York: New York Review Books, 2001, p. 105.
6. Wallace, p. 138.

7. Tom Wolfe. *Hooking Up*. New York: Farrar, Straus and Giroux, 2000, p. 97.
8. Werner Haftmann. *Painting in the Twentieth Century*. New York: Frederick A. Praeger, 1960, pp. 11–12.

## Chapter 26

1. David Salle and Peter Schjeldahl. *An Interview with David Salle by Peter Schjeldahl*. New York: Random House, 1987, p. 48.
2. Kirk Varnedoe. *A Fine Disregard: What Makes Modern Art Modern*. New York: Harry N. Abrams, Inc., 1990, p. 25.
3. Clement Greenberg. *Homemade Esthetics: Observations on Art and Taste*. New York: Oxford University Press, 1999.
4. Susanne K. Langer. *Feeling and Form: A Theory of Art*. New York: Charles Scribner's Sons, 1953.
5. Suzanne K. Langer. *Mind: An Essay on Human Feeling*. 3 vols. Baltimore, Md.: Johns Hopkins University Press, 1967, 1972, 1982.
6. Leo Steinberg. *Other Criteria: Confrontations with Twentieth Century Art*. New York: Oxford University Press, 1972, p. 10.
7. Michael Crichton. *Jasper Johns*. New York: Harry N. Abrams, Inc., 1977, p. 61.

## Chapter 27

1. Brandon Taylor. *Modernism, Postmodernism, Realism*. Hampshire, UK: Winchester School of Art Press, 1987, p. 40.
2. Sam Hunter. *Robert Rauschenberg*. New York: Rizzoli International Publications, Inc., 1999, p. 90.
3. Susanne K. Langer. *Feeling and Form: A Theory of Art*. New York: Charles Scribner's Sons, 1953, p. 331.
4. Paul Wood, Francis Frascina, Jonathan Harris & Charles Harrison. *Modernism in Dispute: Art Since the Forties*. New Haven, Conn.: Yale University Press, 1993, p. 185.
5. Hunter, p. 83.
6. Hunter, p. 120.
7. Hunter, p. 34.
8. Hunter, p. 7.

## Chapter 28

1. Lawrence Alloway. *Roy Lichtenstein*. New York: Abbeville Press, 1983, p. 40.
2. Christopher Knight, *Last Chance for Eden: Selected Art Criticism, 1979–1994*. Edited by Malin Wilson. Los Angeles: Art Issues Press, 1995, p. 129.
3. Alloway, p. 10.
4. Arthur C. Danto. *The Transfiguration of the Commonplace*. Cambridge, Mass.: Harvard University Press, 1981, p. 110.
5. Dave Hickey. *Roy Lichtenstein: Brushstrokes, Four Decades*. New York: Mitchell-Innes & Nash, 2002, p. 15.
6. Alloway, p. 39.
7. Alloway, p. 20.

## Chapter 29

1. Eric Shanes. *Warhol*. London: Studio Editions, 1993, p. 16.
2. Shanes, p. 17.
3. Shanes, p. 22.
4. Arthur C. Danto. *The Philosophical Disenfranchisement of Art*. New York: Columbia University Press, 1988, p. 37.
5. Carter Ratcliff. *The Fate of a Gesture: Jackson Pollock and Postwar American Art*. New York: Farrar, Straus and Giroux, 1996, p. 203.
6. Shanes, p. 20.
7. Shanes, p. 41.
8. Ratcliff, p. 192.
9. John Coplans. "The New Painting of Common Objects," in *Pop Art: A Critical History*, edited by Steven Henry Madoff. Berkeley: University of California Press, 1997, p. 43.
10. Henry Geldzahler. "A Symposium on Pop Art." In *Pop Art*, p. 65.
11. Allen R. Solomon. "The New Art." In *Pop Art*, p. 90.
12. James H. Rubin. *Impressionism*. London: Phaidon Press Limited, 1999, p. 29.
13. Liz Dawtrey, Toby Jackson, Mary Masterton, Pam Meecham & Paul Wood, eds. *Investigating Modern Art*. New Haven, Conn.: Yale University Press, 1996, p. 134.
14. G. R. Swenson. "What Is Pop Art?" In *Pop Art*, p. 103.
15. Shanes, p. 20.

## Chapter 30

1. Bart Kosko. *Heaven in a Chip*. New York: Random House, 2000.
2. Francis Fukuyama. *Our Posthuman*

*Future: Consequences of the Biotechnology Revolution.* New York: Farrar, Straus and Giroux, 2002.

3. Brenda Denzler. *The Lure of the Edge: Scientific Passions, Religious Beliefs, and the Pursuit of UFOs.* Berkeley: University of California Press, 2001.

4. Michael Cuneo. *American Exorcism.* New York: Doubleday, 2001.

5. Irving Sandler. *Art of the Postmodern Era: From the Late 1960s to the Early 1990s.* New York: HarperCollins Publishers, Inc., 1996, p. 223.

## Chapter 31

1. Arthur C. Danto. *After the End of Art.* Princeton, N.J.: Princeton University Press, 1997, p. 167.

2. Danto, p. 170.

3. Julian Bell. *What Is Painting?* New York: Thames & Hudson, Inc., 1999, p. 199.

4. David Hopkins. *After Modern Art.* New York: Oxford University Press, 2000, p. 131.

5. (No author given.) *Julian Schnabel.* The Tate Gallery Publications Department, 1982, p. 10–11.

6. Irving Sandler. *Art of the Postmodern Era: From the Late 1960s to the Early 1990s.* New York: HarperCollins Publishers, Inc., p. 230.

7. *Julian Schnabel*, 1982, pp. 11–13.

8. Gabriella De Ferrari. *Julian Schnabel.* New York: The Pace Gallery, 1992, n.p.

9. De Ferrari, 1992, n.p.

10. Sandler (quoting Thomas Larson), p. 233.

## Chapter 32

1. Matthew Collings. *Blimey! From Bohemia to Britpop: The London Artworld from Francis Bacon to Damien Hirst.* Cambridge, UK: 21 Publishing Ltd., 1998, p. 13.

2. Damien Hirst. *i want to spend the rest of my life everywhere, with everyone, one to one, always, forever, now.* New York: The Monacelli Press, 1997, p. 7.

3. Collings, p. 18.

4. Collings, p. 23.

5. Hirst, p. 10.

6. Erika Langmuir & Norbert Lynton. *The Yale Dictionary of Art and Artists.* New Haven, Conn.: Yale University Press, 2000.

7. Herbert Read and Nikos Stangos, eds. *The Thames and Hudson Dictionary of Art and Artists.* New York: Thames & Hudson, Inc., 1994.

8. Editors of Phaidon Press. *The 20th Century Artbook.* London: Phaidon Press, Ltd., 1996.

9. Jay Tobler, ed. *Art USA.* New York: Phaidon Press, Ltd., 1999.

10. Susan Orlean. "Art for Everybody." *The New Yorker*, Oct. 15, 2001, p. 125.

11. Susanne K. Langer. *Mind: An Essay on Human Feeling, Vol. 1.* Baltimore, Md.: Johns Hopkins University Press, 1967, p. 124.

12. William H. Libaw. *How We Got to Be Human: Subjective Minds with Objective Bodies.* Amherst, N.Y.: Prometheus Books, 2000.

## Chapter 33

1. David Salle & Peter Schjeldahl. *An Interview with David Salle by Peter Schjeldahl.* New York: Random House, 1987, p. 48.

2. Anne Rorimer. "Gerhard Richter: The Illusion and Reality of Painting." *Gerhard Richter: Paintings.* New York: Marian Goodman Gallery, 1987, n.p.

3. Michael Danoff. "Heterogeneity: An Introduction to the Work of Gerhard Richter." *Gerhard Richter: Paintings.* New York: Thames & Hudson, 1988, p. 9.

4. Rorimer, n.p.

5. Robert Storr. *Gerhard Richter October 18 1977.* New York: The Museum of Modern Art, 2000, p. 95.

6. Storr, p. 95.

## Chapter 34

1. Mark Rosenthal. *Anselm Kiefer.* New York: The Art Institute of Chicago and the Philadelphia Museum of Art, 1987, p. 17.

2. John C. Gilmour. *Fire on the Earth: Anselm Kiefer and the Postmodern World.* Philadelphia: Temple University Press, 1990, p. 103.

3. A. G. Cairns-Smith. *Evolving the Mind: On the Nature of Matter and the Origin of Consciousness.* New York: Cambridge University Press, 1996, p. 249.

4. Rosenthal, p. 26.
5. Gerald M. Edelman. *Neural Darwinism: The Theory of Neuronal Group Selection.* New York: Basic Books, 1987.
6. Andrew Newberg, Eugene d'Aquili & Vince Rause. *Why God Won't Go Away: Brain Science and the Biology of Belief.* New York: Ballantine Books, 2001.

## Chapter 35

1. Christopher Knight. *Last Chance for Eden: Selected Art Criticism, 1979–1994.* Ed. Malin Wilson. Los Angeles: Art Issues Press, 1995, pp. 54–55.
2. Peter Scheldahl, in David Whitney, ed. *Eric Fischl.* New York: Stewart, Tabori & Chang, 1988, p. 11.
3. Irving Sandler. *Art of the Postmodern Era: From the Late 1960s to the Early 1990s.* New York: HarperCollins Publishers, Inc., 1996, p. 240.
4. Scheldahl, in Whitney, p. 23.
5. Scheldahl, in Whitney, p. 20.
6. Arthur C. Danto, Robert Enright & Steve Martin. *Eric Fischl: 1970–2000.* New York: The Monacelli Press, 2000, p. 284.

## Chapter 36

1. Peter Schjeldahl. *The Hydrogen Jukebox.* Berkeley: University of California Press, 1991, p. 308.
2. Jeff Koons. *The Jeff Koons Handbook.* New York: Rizzoli International Publications, Inc., 1992, pp. 23–25.
3. Robert Rosenblum. *On Modern American Art.* New York: Harry N. Abrams, Inc., 1999, p. 337.
4. Christopher Knight. *Last Chance for Eden: Selected Art Criticism, 1979–1994.* Ed. Malin Wilson. Los Angeles: Art Issues Press, 1995, p. 112.
5. James Gardner. *Culture or Trash.* New York: Carol Publishing Group, 1993, pp. 208–9.
6. (No author given.) *Jeff Koons.* San Francisco: San Francisco Museum of Modern Art, 1992, p. 109.
7. Koons, pp. 12–13.

## Chapter 37

1. Lisa Phillips. "His Equivocal Touch in the Vicinity of History." In Janet Kardon, *David Salle.* Philadelphia: Institute of Contemporary Art, University of Pennsylvania, 1986, p. 25.
2. Phillips, in Kardon, p. 29.
3. David Salle & Peter Schjeldahl. *An Interview with David Salle by Peter Schjeldahl.* New York: Random House, 1987, p. 39.
4. Salle & Schjeldahl, p. 48.
5. Janet Kardon, "The Old, the New, and the Different," in Kardon, p. 11.
6. Salle & Schjeldahl, p. 36.
7. Robin Dunbar. *The Trouble with Science.* Cambridge, Mass.: Harvard University Press, 1995, p. 6.
8. Edward O. Wilson. *Consilience: The Unity of Knowledge.* New York: Vintage Books, 1998, p. 43.
9. Paul R. Gross and Norman Levitt. *Higher Superstition.* Baltimore, Md.: Johns Hopkins University Press, 1994, p. 72.
10. Edward Lucie-Smith. *Movements in Art Since 1945.* New York: Thames & Hudson, Inc., 1998, p. 226.

## Conclusions

1. Matt Ridley. *The Origins of Virtue.* New York: Viking Penguin, 1997, p. 199.
2. Kristine Stiles & Peter Selz, editors. *Theories and Documents of Contemporary Art.* Berkeley: University of California Press, 1996, p. 168.

# Bibliography

Adler, Kathleen. "The Spaces of Everyday Life." In *Perspectives on Morisot.* T.J. Edelstein, ed. New York: Hudson Hills Press, 1990.
Albee, Edward. "Considering Krasner." In *Lee Krasner.* Los Angeles: Tasende Gallery, 1998.
Alloway, Lawrence. *Roy Lichtenstein.* New York: Abbeville Press, 1983.
Armstrong, Karen. *The Battle for God.* New York: Alfred A. Knopf, 2000.
Auping, Michael, ed. *Abstract Expressionism: The Critical Developments.* New York: Harry N. Abrams, Inc., 1987.
Balmer, Randall. *Mine Eyes Have Seen the Glory.* New York: Oxford University Press, 1989.
Barbour, Julian. *The End of Time: The Next Revolution in Physics.* New York: Oxford University Press, 2000.
Barkow, Jerome H., Leda Cosmides and John Tooby. *The Adapted Mind: Evolutionary Psychology and the Generation of Culture.* New York: Oxford University Press, 1992.
Bartholomew, Robert E., and George S. Howard. *UFOs & Alien Contact: Two Centuries of Mystery.* Amherst, N.Y.: Prometheus Books, 1998.
Barzun, Jacques. *The Use and Abuse of Art.* Princeton, N.J.: Princeton University Press, 1974.
_____. *From Dawn to Decadence.* New York: HarperCollins Publishers, Inc., 2000.
Bell, Julian. *What Is Painting?* New York: Thames & Hudson, Inc., 1999.
Breslin, E.B. *Mark Rothko: A Biography.* Chicago: University of Chicago Press, 1993.
Bronowski, J. *The Visionary Eye.* Cambridge, Mass.: MIT Press, 1978.
Broude, Norma. *Impressionism: A Feminist Reading.* New York: Rizzoli International Publications, Inc., 1991.
Bucke, Richard Maurice, M.D. *Cosmic Consciousness.* New York: E.P. Dutton, 1969.
Bullough, Vern L., and Bonnie Bullough. *Sexual Attitudes.* Amherst, N.Y.: Prometheus Books, 1995.
Cairns-Smith, A.G. *Evolving the Mind: On the Nature of Matter and the Origin of Consciousness.* New York: Cambridge University Press, 1996.
Callen, Anthea. *Spectacular Body.* New Haven, Conn.: Yale University Press, 1995.
Carlyle, Thomas. *Sartor Resartus* (The Tailor Patched). London: J.M. Dent & Sons, introduction by W.H. Hudson, last reprinted 1959.

Cerf, Christopher, and Victor Navasky. *The Experts Speak*. New York: Villard Books, 1998.
Chicago, Judy, and Edward Lucie-Smith. *Women and Art: Contested Territory*. New York: Watson-Guptill Publications, 1999.
Clark, Kenneth. *The Nude*. New York: Pantheon Books, 1956.
_____. *Moments of Vision and Other Essays*. New York: Harper and Row, 1981.
Coles, Robert. *The Secular Mind*. Princeton, N.J.: Princeton University Press, 1999.
Collings, Matthew. *Blimey! From Bohemia to Britpop: The London Artworld from Francis Bacon to Damien Hirst*. Cambridge, UK: 21 Publishing Ltd., 1998.
_____. *This Is Modern Art*. New York: Watson-Guptill Publications, 2000.
Crespelle, Jean-Paul. *Monet: The Masterworks*. New York: Portland House, 1988.
Crews, Frederick. *Skeptical Engagements*. New York: Oxford University Press, 1986.
Crichton, Michael. *Jasper Johns*. New York: Harry N. Abrams, Inc., 1977.
Crow, Thomas. *Modern Art in the Common Culture*. New Haven, Conn.: Yale University Press, 1996.
Danto, Arthur C. *After the End of Art*. Princeton, N.J.: Princeton University Press, 1997.
_____. *The Philosophical Disenfranchisement of Art*. New York: Columbia University Press, 1988.
_____. *The State of the Art*. New York: Prentice-Hall, 1987.
_____. *The Transfiguration of the Commonplace*. Cambridge, Mass.: Harvard University Press, 1981.
_____, Robert Enright, and Steve Martin. *Eric Fischl: 1970–2000*. New York: The Monacelli Press, 2000.
Davies, Hugh. *Bacon*. New York: Abbeville Press, Inc., 1986.
Dawtrey, Liz, Toby Jackson, Mary Masterton, Pam Meecham and Paul Wood, eds. *Investigating Modern Art*. New Haven, Conn.: Yale University Press, 1996.
De Ferrari, Gabriella. *Julian Schnabel*. New York: The Pace Gallery, 1992.
Denvir, Bernard. *Post-Impressionism*. New York: Thames and Hudson, Inc., 1992.
Dijkstra, Bram. *Idols of Perversity: Fantasies of Feminine Evil in Fin-de-Siècle Culture*. New York: Oxford University Press, 1986.
Donoghue, Denis. *The Arts Without Mystery*. Boston: Little, Brown and Company, 1983.
Dunbar, Robin. *The Trouble with Science*. Cambridge, Mass.: Harvard University Press, 1995.
Dunlop, Ian. *Degas*. New York: Galley Press, 1979.
Dunstan, Bernard. *Painting Methods of the Impressionists*. New York: Watson-Guptill Publications, 1983.
Edelman, Gerald M. *Neural Darwinism: The Theory of Neuronal Group Selection*. New York: Basic Books, 1987.
_____, and Giulio Tononi. *A Universe of Consciousness: How Matter Becomes Imagination*. New York: Basic Books, 2000.
Edelstein, T.J., ed. *Perspectives on Morisot*. New York: Hudson Hills Press, 1990.
Eisenman, Stephen F. *Gauguin's Skirt*. New York: Thames and Hudson, Inc., 1997.
Elderfield, John. *The "Wild Beasts": Fauvism and Its Affinities*. New York: Oxford University Press, 1976.
Farr, Dennis. *Francis Bacon: A Retrospective*. New York: Harry N. Abrams, Inc., 1999.
Farwell, Beatrice. "Manet, Morisot, and Propriety." In *Perspectives on Morisot*. T.J. Edelstein, ed. New York: Hudson Hills Press, 1990.
Fer, Briony, David Batchelor and Paul Wood. *Realism, Rationalism, Surrealism*. New Haven, Conn.: Yale University Press, 1993.
Ferloni, Michel, and Dominique Spiess, eds. *Pierre-Auguste Renoir*. New York: Konecky and Konecky, 1994.

Flam, Jack. *Matisse on Art*. Berkeley: University of California Press, 1995.
Flaum, Eric, with David Pandy. *The Encyclopedia of Mythology*. Philadelphia: Running Press Book Publishers, 1993.
Flowers, R. Barri. *The Prostitution of Women and Girls*. Jefferson, N. C.: McFarland and Company, 1998.
Foucault, Michel. *The History of Sexuality*. New York: Pantheon Books, 1985.
_____. *The Use of Pleasure*. New York: Pantheon Books, 1985.
Frascina, Francis, Nigel Blake, Briony Fer, Tamar Garb and Charles Harrison. *Modernity and Modernism: French Painting in the Nineteenth Century*. New Haven, Conn.: Yale University Press, 1993.
Freeman, Julian. *Art: A Crash Course*. New York: Watson-Guptill Publications, 1998.
Fried, Michael. *Art and Objecthood*. Chicago: University of Chicago Press, 1998.
Fry, Edward F. *Cubism*. New York: Oxford University Press, 1978.
Gablik, Suzi. *Has Modernism Failed?* New York: Thames and Hudson, Inc., 1984.
Gardner, Howard. *Art, Mind, and Brain*. New York: Basic Books, 1982.
Gardner, James. *Culture or Trash? A Provocative View of Contemporary Painting, Sculpture, and Other Costly Commodities*. New York: Carol Publishing Group, 1993.
Garvin, Lucius. "The Paradox of Aesthetic Meaning." From *Philosophy and Phenomenological Research (The Journal of Phenomenology)*, Vol. 8 No. 1 (1947), pp. 99–106.
Gaunt, William. *The Impressionists*. New York: Weathervane Books, 1970.
Gay, Peter. *Education of the Senses*. New York: Oxford University Press, 1984.
_____. *The Naked Heart*. New York: W.W. Norton and Company, 1995.
_____. *Pleasure Wars*. New York: W.W. Norton and Company, 1998.
Geertz, Clifford. *Works and Lives: The Anthropologist as Author*. Stanford, Calif.: Stanford University Press, 1988.
_____. *Available Light: Anthropological Reflections on Philosophical Topics*. Princeton, N.J.: Princeton University Press, 2000.
Gellner, Ernest. *Anthropology and Politics: Revolution in the Sacred Grove*. Cambridge, Mass.: Blackwell Publishers Inc., 1995.
*Gerhard Richter: Paintings*. New York: Marian Goodman Gallery, 1987; Thames and Hudson, 1988.
Gilbert-Rolfe, Jeremy. *Beauty and the Contemporary Sublime*. New York: Allworth Press, 1999.
Gilmour, John C. *Fire on the Earth: Anselm Kiefer and the Postmodern World*. Philadelphia: Temple University Press, 1990.
Godfrey, Tony. *Conceptual Art*. London: Phaidon Press Limited, 1998.
Golding, John. *Cubism*. Cambridge, Mass.: Harvard University Press, 1988.
_____. *Paths to the Absolute*. Princeton, N.J.: Princeton University Press, 2000.
Goldwater, Marge. *Past/Imperfect*. Minneapolis: Walker Art Center, 1987.
Gombrich, E.H. *Art and Illusion*. Princeton, N.J.: Princeton University Press, 1989.
Goodman, Nelson. *Languages of Art*. Indianapolis: Hackett Publishing Company, Inc., 1976.
Gould, Stephen Jay. *Rocks of Ages: Science and Religion in the Fullness of Life*. New York: Random House, 2002.
Greenberg, Clement. *Homemade Esthetics: Observations on Art and Taste*. New York: Oxford University Press, 1999.
Gritsch, Eric W. *Born Againism: Perspectives on a Movement*. Philadelphia: Fortress Press, 1982.
Grosenick, Uda, and Burkhard Reimschneider, eds. *Art at the Turn of the Millennium*. New York: Taschen, 1999.

Gross, Paul R., and Norman Levitt. *Higher Superstition*. Baltimore, Md.: Johns Hopkins University Press, 1994.

Haftmann, Werner. *Painting in the Twentieth Century*. New York: Frederick A. Praeger, 1960.

Hagen, Rose-Marie, and Rainer Hagen. *Masterpieces in Detail: What Great Paintings Say*. New York: Taschen, 2000.

Handlin, Oscar. "Science and Technology in Popular Culture." In *Science and Culture*. Gerald Holton, ed. Cambridge, Mass.: Houghton Mifflin Company, 1965.

Hanson, Anne Taylor. *Manet and the Modern Tradition*. New Haven, Conn.: Yale University Press, 1977.

Hanson, Lawrence. *Renoir: The Man, the Painter, and His World*. New York: Dodd, Mead and Co., 1968.

Harris, Marvin. *Our Kind*. New York: Harper and Row, 1989.

Harrison, Charles, Francis Frascina and Gill Perry. *Primitivism, Cubism, Abstraction: The Early Twentieth Century*. New Haven, Conn.: Yale University Press, 1993.

Hickey, Dave. *Roy Lichtenstein: Brushstrokes, Four Decades*. New York: Mitchell-Innes and Nash, 2002.

Hirst, Damien. *i want to spend the rest of my life everywhere, with everyone, one to one, always, forever, now*. New York: The Monacelli Press, 1997.

Hobbs, Robert. *Edward Hopper*. New York: Harry N. Abrams, Inc., 1987.

Hoffman, Donald D. *Visual Intelligence: How We Create What We See*. New York: W.W. Norton, 1998.

Holton, Gerald, ed. *Science and Culture*. Cambridge, Mass.: Houghton Mifflin Company, 1965.

Hopkins, David. *After Modern Art*. New York: Oxford University Press, 2000.

House, John. *Monet: Nature into Art*. New Haven, Conn.: Yale University Press, 1986.

Hughes, H. Stuart. *Oswald Spengler: A Critical Estimate*. New York: Charles Scribner's Sons, 1952.

Hughes, Robert. *Nothing If Not Critical*. New York: Alfred A. Knopf, 1990.

_____. *Culture of Complaint*. New York: Oxford University Press, 1993.

_____. *The Shock of the New*. New York: Alfred A. Knopf, 1996.

Huisman, Philippe. *Morisot: Enchantment*. Translated by Diana Imber. New York: French and European Publications, Inc., 1963.

Hunter, Sam. *Robert Rauschenberg*. New York: Rizzoli International Publications, Inc., 1999.

Huyghe, Rene. *Van Gogh*. New York: Crown Publishers, Inc., 1988.

Huyssen, Andreas. *After the Great Divide*. Bloomington: Indiana University Press, 1986.

*Jasper Johns: Paintings, Drawings and Sculpture, 1954–1964*. London: Whitechapel Gallery, 1964.

James, William. *The Varieties of Religious Experience*. New York: Random House, 1929.

Jobert, Barthelemy. *Delacroix*. Princeton, N.J.: Princeton University Press, 1997.

Johnson, Ellen R., ed. *American Artists on Art: From 1940 to 1980*. New York: Harper and Row, 1982.

*Julian Schnabel*. The Tate Gallery Publications Department, 1982.

Kardon, Janet. *David Salle*. Philadelphia: Institute of Contemporary Art, University of Pennsylvania, 1986.

Knight, Christopher. *Last Chance for Eden: Selected Art Criticism, 1979–1994*. Ed. Malin Wilson. Los Angeles: Art Issues Press, 1995.

Knight, David. *The Age of Science: The Scientific World-view in the Nineteenth Century*. New York: Basil Blackwell Inc., 1986.

Koons, Jeff. *The Jeff Koons Handbook*. New York: Rizzoli International Publications, 1992.
Kosko, Bart. *Heaven in a Chip*. New York: Random House, 2000.
Kranzfelder, Ivo. *Edward Hopper*. New York: Taschen, 1998.
Kurtz, Paul, ed. *Science and Religion: Are They Compatible?* Amherst, N.Y.: Prometheus Books, 2003.
Langer, Susanne K. *Feeling and Form: A Theory of Art*. New York: Charles Scribner's Sons, 1953.
_____. *Mind: An Essay on Human Feeling, Vol. 1*. Baltimore, Md.: Johns Hopkins University Press, 1967.
_____. *Problems of Art*. New York: Charles Scribner's Sons, 1957.
_____, ed. *Reflections on Art*. New York: Oxford University Press, 1961.
Langmuir, Erika, and Norbert Lynton. *The Yale Dictionary of Art and Artists*. New Haven, Conn.: Yale University Press, 2000.
Leight, Hermann. *History of the World's Art*. London: George Allen and Unwin, Ltd., 1952.
Lewis, Bernard. *Islam in History*. Chicago: Open Court, 1993.
Lewontin, Richard. *It Ain't Necessarily So: The Dream of the Human Genome and Other Illusions*. New York: New York Review Books, 2001.
Libaw, William H. *How We Got to Be Human: Subjective Minds with Objective Bodies*. Amherst, N.Y.: Prometheus Books, 2000.
Lipton, Eunice. *Looking into Degas*. Berkeley: University of California Press, 1986.
Lucie-Smith, Edward. *Sexuality in Western Art*. New York: Thames and Hudson, Inc., 1991.
_____. *Movements in Art Since 1945*. New York: Thames and Hudson, Inc., 1995.
_____. *ArtToday*. London: Phaidon Press Limited, 1995.
_____. *Visual Arts in the Twentieth Century*. New York: Harry N. Abrams, Inc., 1997.
Lynton, Norbert. "Expressionism." In *Concepts of Modern Art*. Ed. Nikos Stangos.
Madoff, Steven Henry, ed. *Pop Art: A Critical History*. Berkeley: University of California Press, 1997.
Mageo, Jeannette Marie, and Alan Howard, eds. *Spirits in Culture, History and Mind*. New York: Routledge, 1996.
Marcoci, Roxana, Diana Murphy and Eve Sinaiko, eds. *New Art*. New York: Harry N. Abrams, Inc., 1997.
Margolis, Joseph. *What, After All, Is a Work of Art?* University Park: Pennsylvania State University Press, 1999.
Mathews, Nancy Mowll. *Paul Gauguine: An Erotic Life*. New Haven, Conn.: Yale University Press, 2001.
Matthers, Clifford N., and Roy Abraham Varghese, eds. *Cosmic Beginnings and Human Ends: Where Science and Religion Meet*. Chicago: Open Court, 1995.
Menand, Louis. *The Metaphysical Club*. New York: Farrar, Straus and Giroux, 2001.
Miller, William Ian. *The Anatomy of Disgust*. Cambridge, Mass.: Harvard University Press, 1997.
Muller, Herbert J. *The Children of Frankenstein*. Bloomington: Indiana University Press, 1970.
Muller, Joseph-Emile. *Fauvism*. New York: Frederick A. Praeger, 1967.
Mumford, Lewis. *The Myth of the Machine: The Pentagon of Power*. New York: Harcourt Brace Jovanovich, 1970.
Murphy, Richard W., and the Editors of Time-Life Books. *The World of Cézanne: 1839–1906*. New York: Time-Life Books, 1972.
Nadeau, Robert, and Menas Kafatos. *The Non-Local Universe: The New Physics and Matters of the Mind*. New York: Oxford University Press, 1999.

Nead, Lynda. *The Female Nude: Art Obscenity and Sexuality*. New York: Routledge, 1992.

Newberg, Andrew, Eugene d'Aquili and Vince Rause. *Why God Won't Go Away: Brain Science and the Biology of Belief*. New York: Ballantine Books, 2001.

Panofsky, Erwin. *Meaning in the Visual Arts*. Chicago: University of Chicago Press, 1982.

Picon, Gaeton. *The Birth of Modern Painting*. New York: Rizzoli International Publications, Inc., 1978.

Polkinghorne, John. *Beyond Science*. New York: Cambridge University Press, 1996.

Pool, Phoebe. *Degas*. New York: Marboro Books, 1963.

Prideaux, Tom. *The World of Delacroix: 1798–1863*. New York: Time Inc., 1966.

Ratcliff, Carter. *The Fate of a Gesture: Jackson Pollock and Postwar American Art*. New York: Farrar, Straus and Giroux, 1996.

Read, Herbert. *A Concise History of Modern Painting*. New York: Thames and Hudson, Inc., 1998.

Reid, Louis Arnaud. "Beauty and Significance." From *Proceedings of the Aristotelian Society*, Vol. 14 (1928–29), pp. 123–53.

Rewald, John. *The History of Impressionism*. New York: The Museum of Modern Art, 1973.

_____. *Cézanne: A Biography*. New York: Harry N. Abrams, Inc., 1986.

Rey, Jean Dominique. *Berthe Morisot*. New York: Crown Publishers, Inc. (undated).

Reyburn, Scott. *The Art of the Impressionists*. North Dighton, Va.: JG Press, 1997.

Rosenblum, Robert. *Cubism and Twentieth-Century Art*. New York: Harry N. Abrams, Inc., 1966.

_____. *On Modern American Art*. New York: Harry N. Abrams, Inc., 1999.

Rosenthal, Mark. *Anselm Keifer*. New York: The Art Institute of Chicago and the Philadelphia Museum of Art, 1987.

Ross, Stephen David, ed. *Art and Its Significance*. Albany: State University of New York Press, 1994.

Rubin, James H. *Impressionism*. London: Phaidon Press Limited, 1999.

Salle, David, and Peter Schjeldahl. *An Interview with David Salle by Peter Schjeldahl*. New York: Random House, 1987.

Sandler, Irving. *Art of the Postmodern Era: From the Late 1960s to the Early 1990s*. New York: HarperCollins Publishers, Inc., 1996.

Schapiro, Meyer. *Impressionism: Reflections and Perceptions*. New York: George Braziller, Inc., 1997.

Schjeldahl, Peter. *The Hydrogen Jukebox: Selected Writings, 1978–1990*. Ed. Malin Wilson. Berkeley: University of California Press, 1991.

_____. "The Urbane Innocent." *The New Yorker*, Nov. 20, 2000.

Selz, Peter. *Beyond the Mainstream: Essays on Modern and Contemporary Art*. New York: Cambridge University Press, 1997.

Shanes, Eric. *Warhol*. London: Studio Editions, 1993.

Shattuck, Roger. *Forbidden Knowledge: From Prometheus to Pornography*. New York: St. Martin's Press, 1996.

Shermer, Michael. *How We Believe*. New York: W.H. Freeman, 1999.

_____. *The Borderlands of Science*. New York: Oxford University Press, 2001.

_____. *Why People Believe Weird Things*. New York: Henry Holt, 2002.

Shore, Bradd. *Culture in Mind: Cognition, Culture, and the Problem of Meaning*. New York: Oxford University Press, 1996.

Silver, Brian L. *The Ascent of Science*. New York: Oxford University Press, 1998.

Singer, Charles. *A Short History of Scientific Ideas to 1900*. New York: Oxford University Press, 1959.

Smith, Huston. *Why Religion Matters: The Fate of the Human Spirit in an Age of Disbelief.* San Francisco: HarperCollins Publishers, Inc., 2002.
Snow, C.P. *The Two Cultures.* New York: Cambridge University Press, 1993.
Solomon-Godeau, Abigail. *Male Trouble: A Crisis in Representation.* New York: Thames and Hudson, Inc., 1997.
Sorrentino, Gilbert. "Kitch into 'Art': The New Realism." In *Pop Art.* Steven Henry Madoff, ed. Berkeley: University of California Press, 1997.
Stangos, Nikos, ed. *Concepts of Modern Art: From Fauvism to Postmodernism.* New York: Thames and Hudson, Inc., 1994.
Steinberg, Leo. "The Eye Is a Part of the Mind." From *Partisan Review*, Vol. 20, No. 2 (1953).
_____. *Other Criteria: Confrontations with Twentieth Century Art.* New York: Oxford University Press, 1972.
Steiner, Wendy. *The Scandal of Pleasure.* Chicago: The University of Chicago Press, 1995.
Stern, Paul. "On the Problem of Artistic Form." From *Logos*, Vol. 5 (1914–15), pp. 165–72.
Stevenson, Leslie, and David L. Haberman. *Ten Theories of Human Nature.* New York: Oxford University Press, 1998.
Stiles, Kristine, and Peter Selz. *Theories and Documents of Contemporary Art.* Berkeley: University of California Press, 1996.
Stoyanov, Yuri. *The Other God: Dualist Religions from Antiquity to the Cathar Heresy.* New Haven, Conn.: Yale University Press, 2000.
Strickland, Carol, and John Boswell. *The Annotated Mona Lisa.* Kansas City, Mo.: Andrews McMeel Publishing, 1992.
Sutton, Denys. *Edgar Degas: Life and Work.* New York: Rizzoli International Publications, Inc., 1986.
Swenson, G.R. "What Is Pop Art?" In *Pop Art.* Steven Henry Madoff, ed. Berkeley: University of California Press, 1997.
Sylvester, David. *About Modern Art.* New York: Henry Holt, 1997.
_____. *Looking Back at Francis Bacon.* New York: Thames and Hudson, Inc., 2000.
Taillandier, Ivan. *Cézanne.* New York: Crown Trade Paperbacks, 1977.
Taylor, Basil. *The Impressionists and Their World.* London: Phoenix House, 1953.
Taylor, Brandon. *Modernism, Post-modernism, Realism: A Critical Perspective for Art.* Hampshire, UK: Winchester School of Art Press, 1987.
Thomson, Belinda. *Gauguin.* New York: Thames and Hudson, Inc., 1987.
_____. *Impressionism: Origins, Practice, Reception.* New York: Thames and Hudson, Inc., 2000.
Tierney, Brian, Donald Kagan and L. Pearce Williams, eds. *Great Issues in Western Civilization, Vol. II*, third edition. New York: Random House, 1976.
Time-Life Books, eds. *American Painting: 1900–1970.* New York: Time-Life Books, 1970.
Tobler, Jay, ed. *Artusa.* New York: Phaidon Press Ltd., 1999.
Trapp, Frank Anderson. *The Attainment of Delacroix.* Baltimore, Md.: Johns Hopkins University Press, 1970.
Updike, John. "The Future of Faith." In *The New Yorker*, 11/29/99.
Varnedoe, Kirk. *A Fine Disregard: What Makes Modern Art Modern.* New York: Harry N. Abrams, Inc., 1990.
Varnedoe, Kirk. *Jasper Johns: A Retrospective.* New York: The Museum of Modern Art, 1996.
Varnedoe, Kirk, and Adam Gopnik. *High and Low: Modern Art/Popular Culture.* New York: The Museum of Modern Art, 1990.
Varnedoe, Kirk, with Pepe Karmel. *Jackson Pollock.* New York: The Museum of Modern Art, 1998.

Vogt, Paul. *Expressionism: German Painting, 1905–1920*. New York: Harry N. Abrams, Inc., 1980.
Wagner, Ann Middleton. *Three Artists*. Berkeley: University of California Press, 1996.
Wallace, B. Alan. *The Taboo of Subjectivity: Toward a New Science of Consciousness*. New York: Oxford University Press, 2000.
Wechsler, Herman J. *French Impressionists*. New York: Harry N. Abrams, Inc. (no date).
Weinstein, Leo, ed. *The Age of Reason: The Culture of the Seventeenth Century*. New York: George Braziller, 1965.
Weintraub, Linda, Arthur Danto and Thomas McEvilley. *Art on the Edge and Over*. Litchfield, Conn.: Art Insights, Inc., 1996.
Weschler, Lawrence. "The Looking Glass." In *The New Yorker*, Jan. 21, 2000, p. 65.
White, Barbara Ehrlich. *Impressionists Side by Side*. New York: Alfred A. Knopf, 1996.
Whitney, David, ed. *Eric Fischl*. New York: Stewart, Tabori and Chang, 1988.
Wilson, A.N. *God's Funeral*. New York: W.W. Norton, 1999.
Wilson, Colin. *Alien Dawn*. New York: Fromm International Publishing Corp., 1988.
Wilson, Edward O. *Consilience: The Unity of Knowledge*. New York: Vintage Books, 1998.
Wolfe, Tom. *Hooking Up*. New York: Farrar, Straus and Giroux, 2000.
Wollheim, Richard. *Art and Its Objects*. New York: Cambridge University Press, 1980.
Wood, Paul, Francis Frascina, Jonathan Harris and Charles Harrison. *Modernism in Dispute: Art Since the Forties*. New Haven, Conn.: Yale University Press, 1993.
Wright, Robert. *The Moral Animal*. New York: Pantheon Books, 1994.
Yenawine, Philip. *How to Look at Modern Art*. New York: Harry N. Abrams, Inc., 1991.

# *Index*

Numbers in ***bold italics*** represent illustrations.